T0321768

Cognitive Computing for Internet of Medical Things

Cognitive Computing for Internet of Medical Things (IoMT) offers a complete assessment of the present scenarios, roles, challenges, technologies, and impacts of IoMT-enabled smart healthcare systems. It contains chapters discussing various biomedical applications under the umbrella of the IoMT.

Key Features

- Exploits the different prospects of cognitive computing techniques for the IoMT and smart healthcare applications
- Addresses the significance of IoMT and cognitive computing in the evolution of intelligent medical systems for biomedical applications
- Describes the different computing techniques of cognitive intelligent systems from a practical point of view: solving common life problems
- Explores the technologies and tools to utilize IoMT for the transformation and growth of healthcare systems
- Focuses on the economic, social, and environmental impacts of IoMT-enabled smart healthcare systems.

This book is primarily aimed at graduates, researchers, and academicians working in the area of development of the applications of IoT in smart healthcare. Industry professionals will also find this book helpful.

Cognitive Computing for Internet of Medical Things

Edited by
A Prasanth
Lakshmi D
Rajesh Kumar Dhanaraj
Balamurugan Balusamy
Sherimon P C

CRC Press
Taylor & Francis Group
Boca Raton London New York

CRC Press is an imprint of the
Taylor & Francis Group, an **informa** business

A CHAPMAN & HALL BOOK

First edition published 2023
by CRC Press
6000 Broken Sound Parkway NW, Suite 300, Boca Raton, FL 33487-2742

and by CRC Press
4 Park Square, Milton Park, Abingdon, Oxon, OX14 4RN

CRC Press is an imprint of Taylor & Francis Group, LLC

© 2023 selection and editorial matter, A Prasanth, Lakshmi D, Rajesh Kumar Dhanaraj, Balamurugan Balusamy and Sherimon P C; individual chapters, the contributors

ISBN: 978-103-218-7884 (hbk)
ISBN: 978-103-218-7891 (pbk)
ISBN: 978-100-325-6243 (ebk)

DOI: 10.1201/9781003256243

Typeset in Palatino
by codeMantra

Contents

Preface

Technology can help us in our personal and professional life in the quest to stay healthy. Our digital devices can help us to improve our diets, measure our workout activities, and keep track of our prescription compliance.

Sensor-based devices such as physiological monitors, ventilators, infusion pumps, and bedside computer terminals are utilized in modern medical treatment for the reading of biomarkers and other clinical activities.

Reducing human errors, increasing clinical outcomes, facilitating care coordination, enhancing practice efficiencies, and collecting data over time are all examples of how health informatics may help improve and revolutionize healthcare. Machine learning, the Internet of things (IoT), deep learning, embedded systems, augmented reality, and cognitive computing are among the technologies used.

The usage of medical technological instruments ensures the safety of patients. First, there are medication warnings, flags and reminders, consultation and diagnosis reports, and improved patient data accessibility. Alerts, in particular, might assist someone in sticking to specified medications and treatment schedules.

Digital technology in medicine and healthcare could help transform unsustainable healthcare systems into sustainable ones, ensure the equal relationships between medical professionals and patients, and provide cheaper, faster, and more effective disease solutions – technologies could help us win the cancer battle.

Doctors and nurses now use mobile gadgets to record real-time data and update patients' medical histories. Diagnoses and treatments become more accurate and efficient as a result. The centralization of key patient data and lab findings has significantly improved healthcare quality.

The use of artificial intelligence-based analytical models to imitate the human thought process in complex settings where the solutions may be ambiguous and uncertain is known as cognitive computing. The goal of cognitive computing is to create a computational model that mimics human thought processes. The computer may simulate the way the human brain works by using self-learning algorithms that leverage data mining, pattern recognition, and natural language processing. Cognitive computing systems are thinking, reasoning, and remembering systems that collaborate with humans to help them make better decisions. Its findings are meant to be consumed by humans. AI aims to produce the most accurate result or action by employing the best algorithm.

Thanks to 5G's 100 times more bandwidth than 4G for connecting with IoT devices, healthcare service providers can rely on remote patient monitoring or wearable devices to continuously gather, report, and transfer crucial information to a remote monitoring station.

IoT has the potential to make healthcare more affordable and efficient in future. It can aid in the development of more personalized and patient-centered devices. Furthermore, IoT will enable patients to have better access to data and individualized care, resulting in fewer hospital visits.

The Internet of Medical Things (IoMT) is a collection of medical equipment and apps that use online computer networks to link to healthcare IT systems. Machine-to-machine communication, which is the foundation of IoMT, is enabled by medical devices equipped with Wi-Fi.

A communication protocol is a set of rules that allows two or more entities in a communications system to send data using any physical quantity variation. The protocol specifies the communication rules, syntax, semantics, and synchronization, as well as error recovery techniques.

IoT protocols are classified according to the role that each of them plays in the network. Protocols for network infrastructure (e.g., 6LowPAN), communications (Wi-Fi and Bluetooth), data transmission (MQTT, CoAP, and XMPP), security (DTLS), device management, and telemetry (LwM2M) are just a few examples.

The Internet of Medical Things (IoMT) is a network of sensors, wearable devices, medical devices, and clinical systems that are all connected.

The embedded medical device takes the user's inputs and compares them to a pre-loaded symptom dossier and then attempts to match the symptoms with the appropriate ailment. If the disease is not detected by evaluating the symptoms, it undertakes tests suggested by the pre-loaded symptom file to find a precise match for the condition.

Embedded systems are also used in glucose monitors, pacemakers, CPAP machines, and a range of biomedical sensors. Embedded systems in biomedical applications allow clinicians to use telemedicine and other remote systems to remotely monitor patients' health and make diagnostic and treatment decisions.

The security of data and devices has become a serious problem in the field of healthcare automation. This book covers a variety of tools and approaches for dealing with the safety and security systems of medical equipment.

This book is a collection of chapters written for health professionals, academicians, undergraduates, graduates, and researchers who want to learn more about the cutting-edge technologies employed in modern drugs.

Editors

Dr. A Prasanth is an Assistant Professor at Sri Venkateswara College of Engineering, Sriperumpudur, India. He earned a B.E. degree in Electronics and Communication Engineering from Anna University, Chennai, and an M.E degree in Computer Science and Engineering (with specialization in networks) from Anna University, Chennai, and also earned a Ph.D. in Information and Communication Engineering from Anna University, Chennai, India. Recently, he has received the Young Scientist Award from International Scientist Awards 2020 for his excellent research performance. He also received the Researcher of the Year Award 2020 from "2nd International Business and Academic Excellence Awards (IBAE)", in Delhi, held on December 26, 2020. Moreover, he received the Young Researcher Award from "Institute of Scholars Awards" 2020. He has published more than 25 research articles in reputed international journals, among which 7 articles are indexed in SCI and 15 articles are indexed in Scopus. He has published and granted two patents. Further, he has published more than eight books under reputed publishers. He has served as a resource person in 20 AICTE-Sponsored STTP programs. Moreover, he has served as an Editorial Board Member in various reputed SCI journals. He has 8 years of teaching experience, and his research interests include Internet of things, edge computing, cloud computing, and 5G network.

Dr. Lakshmi D is a Senior Associate Professor (Grade 2) in the School of Computer Science and Engineering at VIT Bhopal University, Madhya Pradesh, India. Since February 2021, she was designated as an Educational Research Officer at Vishnu Educational Development and Innovation Centre (VEDIC) and an Associate Professor at B V Raju Institute of Technology run by Shri Vishnu Educational Society, Hyderabad, from 2016 to Feb 2021. She has been working in the educational sector since 1998. Her key focus is on exploring the dynamics of learning, dynamics of learner, and classroom dynamics, suitable to accelerate the learning efficacy of higher education students. With her expertise, she has delivered more than 150 sessions on various titles. Her research areas include machine learning, deep learning, Internet of things, educational technology, educational data mining, virtual education, and educational psychology. She holds a Ph.D. degree in Information and Communication Engineering from Anna University, Chennai, India. She has published 28 articles in various journals and conference proceedings and contributed chapters to books. She has won two best paper awards: one in IEEE and one in Springer conference. One book publication titled, *"Theory of Computation"*, seven Indian patents provisionally published and waiting for examinations and filed one copyright waiting for grant. One copyright has been granted, two Australian patents have been granted, and one is filed and waiting for a grant.

Dr. Rajesh Kumar Dhanaraj is an Associate Professor in the School of Computing Science and Engineering at Galgotias University, Greater Noida, Uttar Pradesh, India. He holds a Ph.D. degree in Information and Communication Engineering from Anna University, India. He has presented papers at conferences, published articles and papers in various journals, and contributed a chapter to a book. His research and publication interests include wireless sensor networks and cloud computing. He is an Expert Advisory Panel Member of Texas Instruments Inc., the USA.

Balamurugan Balusamy is currently an associate dean student at Shiv Nadar University, Delhi-NCR. Prior to that he was a professor at the School of Computing Sciences & Engineering and Director of International Relations at Galgotias University, Greater Noida, India. His contributions focus on Engineering Education, Block chain and Data Sciences. His Academic degrees and twelve years of experience working as a faculty member in a global University, VIT University, Vellore has made him more receptive and prominent in his domain. He has 200 plus high impact factor papers in Springer, Elsevier and IEEE. He has done more than 80 edited and authored books and collaborated with eminent professors across the world from top QS ranked universities. Balusamy has served up to the position of associate professor in his 12 years with VIT University, Vellore.

He had completed his Bachelors, Masters and PhD Degrees from top premier institutions in India. His passion is teaching and he adapts different design thinking principles while delivering his lectures. He has published 80+ books on various technologies and visited 15 plus countries for his technical course. He has several top-notch conferences in his resume and has published over 200 quality journal articles, conference papers and book chapters combined. He serves on the advisory committee of several startups and forums and does consultancy work for industry on Industrial IOT. He has also given over 195 talks at various events and symposiums.

Dr. Sherimon P C is presently working as a Faculty of Computer Studies at the Arab Open University, Oman. He has 22 years of international experience in administration, teaching, and research. He has completed numerous research projects and consultancy projects. His Google Scholar citation is 316, h-index is 7, and i10 index is 6.

List of Contributors

P. Anbarasu
Sri Eshwar College of Engineering
Coimbatore, India

Pon Bharathi A
Amrita College of Engineering and
    Technology
Nagercoil, India

Rajesh Kumar Dhanaraj
Galgotias University
Greater Noida, India

Arunkumar Gopu
VIT-AP University
Andhra Pradesh, India

V. Karuppuchamy
Kongunadu College of Engineering and
    Technology,
Tholurpatti, India

M. Kavitha
Sri Krishna College of Engineering and
    Technology
Coimbatore, India

Ashish Kumbhare
The ICFAI University
Raipur, India

Lakshmi D
VIT Bhopal
Madhya Pradesh, India

S. Lavanya
Muthayammal Engineering College
Rasipuram, India

Murugan Mahalingam
SRM Valliammai Engineering College
Chennai, India

C. Manikandan
Pandian Saraswathi Yadav Engineering
    College
Sivagangai, India

M. Nalini
Sri Sairam Engineering College
Chennai, India

S. Palanivel Rajan
M. Kumarasamy College of Engineering
Karur, India

A. Prasanth
Sri Venkateswara College of Engineering
Sriperumbudur, India

N. Pushpalatha
Sri Eshwar College of Engineering
Coimbatore, India

Revathi Arumugam Rajendran
SRM Valliammai Engineering College
Chennai, India

Preethi Sambandam Raju
SRM Valliammai Engineering College
Chennai, India

S. Roobini
SNS College of Technology
Coimbatore, India

C. Soundaryaveni
Sri Krishna College of Engineering and
    Technology
Coimbatore, India

K.K. Devi Sowndarya
DMI College of Engineering
Chennai, India

M. Sujaritha
Sri Krishna College of Engineering and
  Technology
Coimbatore, India

S. Swathi
Sri Sairam Institute of Technology
Chennai, India

Piyush Kumar Thakur
The ICFAI University
Raipur, India

S. Veluchamy
Sri Venkateswara College of Engineering
Sriperumbudur, India

Neelanarayanan Venkataraman
VIT University, Chennai,
India

A. Venkatesh
Dr. Mahalingam College of Engineering
  and Technology
Coimbatore, India

S. Vijayanand
Sri Venkateswara College of Engineering
Sriperumbudur, India

Allan J Wilson
Amrita College of Engineering and
  Technology
Nagercoil, India

1

Toward the Internet of Things and Its Applications: A Review on Recent Innovations and Challenges

Arunkumar Gopu
VIT-AP University

Neelanarayanan Venkataraman
VIT University, Chennai

M. Nalini
Sri Sairam Engineering College

CONTENTS

1.1 Introduction

Internet of things (IoT) is an emerging paradigm to connect the sensing units with a processing powerhouse using a reliable communication network [1]. The sensor itself is an autonomous unit capable of observing environmental events. The digital or analog signals are passed to processing units to attain actionable knowledge. IoT is an abstract term for "connecting the unconnected entities". IoT covers broad technological aspects from sensing hardware components to protocols implemented in heterogeneous devices for

DOI: 10.1201/9781003256243-1

interoperable communication. The uncountable number of diversified devices connected to the network made IoT an emerging and active research area [2].

IoT began its journey in late 2008, and it has seen significant advancement in most of the fields. During its inception, the number of devices connected to the network was very less. In the present scenario, 99% of electrical and electronic devices remain unconnected. As the connectivity has matured, the network can support a greater number of heterogeneous devices, but the devices are still obsolete. The remaining 1% of devices connected to the network produce massive amounts of structured and unstructured data. The key factor driving the IoT ecosystem is digitization.

The non-automated human intervention systems are now replaced with sophisticated sensor interactions. The sensory data are collected, shifted, and applied to give rise to a data-driven decision-making process. For example, the electrical meters in Indian households are analogous. In the year 2015, the government decided to roll out the smart metering where the readings are automatically queried using a GPRS connection to generate electricity bills. At present, the project is still in the implementation stage in most states. The implementation time outdated the technology used in smart metering. The predominant IoT issues are scaling, legacy device support, security, data storage and retrieval, and real-time data analytics.

The organization of our chapter is as follows. The below heading introduces the layered IoT architecture to cover the basic understanding of the IoT ecosystem. In the next section, we introduce IoT usage in various application fields and its associated challenges.

1.1.1 Sensor Layer

The bottommost layer in an IoT architecture is the sensor layer as depicted in Figure 1.1. This layer groups all the endpoint devices that sense the environment to generate the data [3]. In this layer, the primary function of the devices is to generate data and efficiently pass them to the base station. Most of the sensor devices do not have processing capabilities. Some sensors are battery-powered and have very little storage and less coverage area. The sensor sends data only when an event is detected. The directly powered sensors don't have such restrictions. Battery-powered sensors are sometimes configured to send relative values instead of absolute values to conserve battery power.

For example, a temperature sensor is implemented to measure the readings of the furnace. If the temperature exceeds the threshold, the sensors are going to send a relative value in terms of "high" to the base instead of sending the absolute temperature reading. The next-generation sensors are called smart objects that have minimal processing units, sensing units, power, and transmission units. Each sensor has the following properties:

1. The signal propagation distance of a sensor defines how long it can transmit the data without data loss at a constant transmission speed.

2. **Frequency bands:** The sensor transmits the data using a licensed spectrum (GPRS) or an unlicensed spectrum (Zigbee). The unlicensed spectrum spans a lesser distance compared to a licensed one.

3. **Topology:** This defines the connectivity layout between devices. A sensor directly connected to a gateway device forms a star topology. The sensors that form star topology are called reduced functionality devices (RFDs). These RFDs are not capable of forwarding the data from other sensors. Fully functional devices (FFD) form a mesh network where each sensor can forward the data of other sensors to reach the base station or sink node.

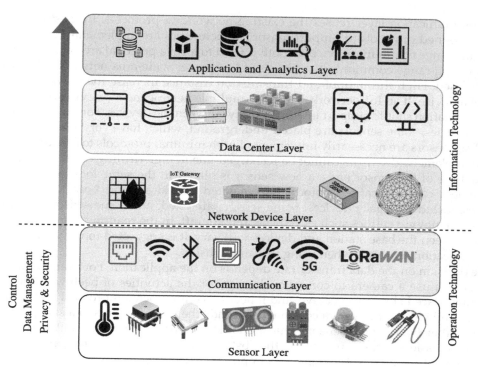

FIGURE 1.1
IoT layered architecture.

Actuators also belong to this layer. The sensor's control signals are converted to actions using this actuator. For example, in an agricultural farm, the temperature sensors measure the reading and pass it to the base station. When the temperature is critically high, it triggers the sprinkler actuator. Most of the sensors have limited capabilities in terms of memory, processing, battery power, communication distance, and transmission speed. The sensor capabilities influence the design of any IoT ecosystem.

1.1.2 Communication Layer

The communication layer connects the devices to the base station for effective data transmission. The factors to be considered when selecting communication protocols are given below with possible options that impact the selection.

- **Device Pool:** heterogeneous devices, homogeneous devices.
- **Device Nature:** static device, roaming device.
- **Power Source:** battery, direct power, sustainable energy.
- **Implementation Space:** air space, underwater, underground.
- **Transfer Type:** continuous transfer, event-triggered transfer, periodic transfer.
- **Transfer Rate:** bits per second.
- Security.
- Tolerance for packet loss.

The communication layer reuses the existing networking protocols. In most cases, specially designed protocols to support sensor characteristics are implemented in this layer. For short-range communication, Bluetooth is used to form a personal area network (PAN), and for long-range communication, licensed low-power wide-area network (LPWAN) is widely used. High throughput and transmission distance increase the power consumption of a device. If an IoT network comprises multiple heterogeneous devices, the protocol implementation should support interoperability between the devices. For example, in mining industries, some sensors are placed underground, which have very limited capabilities. The sensors are necessarily implemented with minimal protocols to conserve energy.

The implemented sensors cannot be recharged due to high drilling costs. Instead, once the lifetime of the sensor ends, a new sensor is placed in the same location. The wired or wireless communication protocol implemented should support both RFD and FFD devices, including legacy devices. Some sensors are fixed, and some sensors are implemented in moving entities such as vehicles and aircraft. In the roaming situation, the distance between the base station and devices will vary. It is suggested to use licensed band communication protocols when using roaming sensors.

The decision on the data transfer rate depends on the application. For example, a group of scientists use a camera to continuously monitor the activities of beehives. The video camera is placed close to a beehive, and it needs a communication protocol that supports higher bandwidth. The camera continuously sends the video streams to a storage device. On the other hand, some sensors send periodic data only when an event is triggered. The transmission distance reduces as the obstacles increase between the communication paths. Security and packet loss are also important factors while choosing a communication protocol.

1.1.3 Network Device Layer

The network device layer comprises edge gateways, switches, and routers. The edge gateway device enables device-to-device communication. Edge gateway supports a variety of devices from different manufacturers implemented with a variety of protocols. The protocol translation methods are implemented in gateways to ensure compatibility in communication between the sender and the receiver devices.

The protocol header is removed by the gateways, and the data are encapsulated to match the devices in the higher layers. In wireless connectivity, a single gateway device can support up to 5,000 sensor nodes. Some other functions performed by an advanced gateway device are data processing, data filtering, and ensuring security. The gateway device presents the data collected from the sensors to networking switches and routers. The switches are routers that work using traditional IPv4 or IPv6 matured protocols.

The IoT network is unimaginably massive and exhausted IPv4 addresses and now IPv6 came into existence. The devices are massively distributed in various geographical locations, leading to a broader security exploit. There are possibilities that a device is compromised to send misleading data to the base stations. Device-level authentication and encryption need to be implemented effectively in all the end devices. This networking device passes the data to the data center layer for further processing.

1.1.4 Data Center Layer

The data center layer consists of all the storage devices and data processing equipment to work on the collected data [4]. The data center has flexibility in implementation within the organization boundary or opted for cloud-based services. The decision on the selection scheme of the data center should also favor the application use cases. For example, a manufacturing industry positioned in a single location collects data in its own data centers. An organizational data center can be accessed by different stakeholders based on their needs. In contrast, if a manufacturing industry spawns in multiple locations, then it is recommended to opt for cloud-based data centers.

The cloud-based data centers enable unified access to all the stakeholders of the industry. Cloud data centers also help in correlating information from a different source in a centralized location. The authentication and authorization techniques limit access based on their roles and responsibilities. A cloud data center is a cost-effective implementation. For example, if the industry is in multiple locations, implementing a data center in each location incurs greater investments in terms of infrastructure procurement cost, manpower to maintain the infrastructure, and maintenance costs. Instead, the cloud model provides the computation and storage infrastructures on-demand in pay-as-you-go model.

Scalability and data mobility are the main advantages of using cloud-based platforms. For example, an aircraft has thousands of sensors implemented in its twin engines that generate TBs of data per hour. To process the collected data, a massive data center is needed. As the number of trips increases, the traditional data centers cannot scale effectively to process the data. Cloud data centers are distributed, they tackle the issue very effectively by offloading the processing to idle infrastructure, and the results are generated in a reasonable time.

1.1.5 Application and Analytics Layer

The application layer consumes the generated data to make business decisions. In the application layer, applications are developed to perform data analytics. The type of structured and unstructured data is cleaned, loaded, and transformed into an actionable format to arrive at better decisions. For example, the beehive video streams are unstructured data. To derive knowledge from the data source, it needs to be processed to check for any malicious events in the locality, such as wasp attacks.

The algorithms to classify the normal behavior of honeybees and their behavior when they are exposed to attacks are implemented in this application layer. Real-time analytics is yet a major challenge that needs to be addressed. The video streams generate a large amount of data that need to be efficiently transferred to the cloud system. It requires uninterrupted higher bandwidth to pass the data to a remote data center. The data center needs to process them immediately to send an alert to the farm owner that a beehive is under attack. Also, the actions need to be performed in real time. To overcome this issue, the edge analytics model is majorly followed. Edge analytics reduces the latency of data transfer. Instead, a device capable of generating alerts is implemented in the lowest possible layers. However, the data are stored in the cloud for future analytics.

Analytics is extensively used with data collection to predict future business trends. For example, a car is fixed with sensors to measure the wear and tear of various types, and the data are sent to the manufacturers to produce the product based on the user's demands. The IoT broker links the needs of the customers to industries that consecutively increase sales. In the present situation, identifying the viable customer to sell the product is highly disorganized. Analytics is a competent technology to bridge the gap between buyers and sellers.

1.2 IoT and Transportation

In view of the ever-increasing population and vehicle-based congestion, IoT-based smart transport is an emerging strategic plan implemented in most of the developed nations to reduce the emission of greenhouse gases. Smart transport has positive impacts on reducing vehicular congestion, increasing economical values, and improving user driving experience. IoT smart transport is an assortment of independent projects that include smart parking, ridesharing, traffic control, and traffic information system. In this chapter, we see how this smart transport solution works, its impact, and the challenges.

1.2.1 Smart Parking

At present, more than half of the world's population lives in urban cities. Cities bring more opportunities to the people. In recent years, the migration of people has increased multifold. The topmost reason for migration is job opportunities and better educational opportunities. But still, the landscape remains unchanged. The streets are becoming even denser with people. New expansions of cities are comparatively less to the number of people migrating to the urban areas. People commute using their vehicles or using public transport, which leads to an increase in congestion.

The cities are not well adapted to the increasing population density. In particular, transport plays an important role in the development of cities. As the number of vehicles increases, it is even harder to find a parking place in major cities where the persons/sq. mile is very high. A study on population statistics shows that New York City has 24,692 people/sq. mile and has the densest population in the entire world with only 18% of parking space [5].

Finding parking in such a crowded city is an unfruitful experience for many people. To solve the issue, efficient off-street parking infrastructures are proposed. Many cities have implemented a real-time parking information system that directs the drivers to the nearest available parking space. An efficient solution helps in reducing driver's search time, environmental pollution, fuel cost, and congestion. With sensor technologies and advancements in image processing, the identification of unauthorized parking increases the city's revenue.

Monitoring parking space availability is a crucial function, and all the remaining actions are performed based on availability. Throughout the literature, there are two types of sensing methods that are implemented to detect space availability. A static sensor is implemented in each of the parking spaces. When a vehicle is parked, the sensor detects and sends the information to the base station with an accuracy of 95% in 60 seconds. But this method needs a greater number of sensors implemented in the parking location.

When a parking space becomes very large, individual sensors placed in the parking location are not economical. The maintenance and communication burden increases as the parking space is employed with many sensors. The second method is based on vehicle detection through its movement. When a vehicle passes through a parking station, the station detects its license plates and vehicle's GPS signals. This method utilizes a lesser number of sensors, but accuracy is comparatively less. Various combinations of sensing devices, namely infrared sensors, ultrasonic sensors, optical sensors, magnetometers, acoustic sensors, RFID, and cameras, are commonly used [6].

Sensors are connected to the network using any one of the connectivity protocols such as Bluetooth/BLE, Zigbee, and 802.11ah. 802.11ah protocol is a specially designed Wi-Fi protocol that works in the sub-GHz band to conserve the energy of battery-powered devices. The prime objectives of parking systems are listed incrementally based on maturity levels.

1. Improved user experience by reducing the driver's search time.
2. Efficient utilization of parking space.
3. Real-time parking information system to direct the driver to the nearest available parking space.
4. Integrated payment system for seamless wireless fund transfer.
5. Reduce pollution and congestion.
6. Real-time data analytics.

Privately owned off-street parking spaces are frequently used during working hours. The private space can also be shared with the public to satisfy the parking needs during off-hours. This has great potential in increasing the city's revenue by efficiently utilizing the available parking space. To direct the drivers to a parking location within the parking lot, digital sign boards are linked with the information system. The digital sign boards change dynamically to direct the drivers to an appropriate parking space.

Challenges and Open Issues

1. Identifying a parked vehicle based on readings from a single sensor often leads to undesirable results. A combination of sensors is used to identify the parked vehicle, which increases the capital and maintenance expenditure. Sensors are once again selected based on environmental conditions. An optical sensor detects light energy and converts it to electrical signals, which cannot be implemented in on-street-based parking solutions. Ensuring secure connectivity is also a challenge considering a fast-moving vehicle. When a vehicle enters a parking location, the connection is initiated. As the vehicle moves faster, it leaves the sensing range of the communicating device. A non-closed connection may lead to security exploits in the vehicle sensing system. Coordinated implementation of sensing device networks reduces such exploits to a certain extent [7].

2. Devices with wider sensing areas and less battery power consumption are the need of the hour. Longer communication and data transfer between devices consume more energy. A city-level parking information system needs multiple sensors integrated for correlation between the collected data [8]. Private parking space typically uses short-range communication using unlicensed bandwidth. Longer communication protocols such as GPRS, 3G, and 4G consume more energy from sensing devices. Even though low-power alternatives are in practice, power-efficient sensor and communication protocol development is still an active research area.

3. Security and privacy is a major issue when dealing with personal information [9]. The information that can be misused by the stakeholders is the flip side of IoT-based solutions. The user location data are continuously streamed for real-time traffic management-based applications. Different application providers need to utilize the data for navigating the drivers to the best possible routes to reach the destination. Protecting consumer data is increasingly difficult considering a larger IoT ecosystem. A compromised sensing device in a network can collect all the

information that passes through the peer-to-peer network. Implementing security measures in end-user devices is a highly challenging task. Effective implementation of encryption algorithms is needed without burdening the limited computational power of the end devices.

4. Framework for consuming the collected data is the best practice considering multiple stakeholders [10]. The user/driver queries the system in real time for traffic analysis or parking slot availability. The user needs a holistic view of the present situation around the area. The vehicle communicates based on data that are available at the moment. A parking-lot manager queries periodic parking records to forecast the near-future demands. An analyst queries a large amount of historical data to make business decisions. A government organization may come up with a proposal to construct a newer parking lot in locations that have the most parking demands or construction of over-bridges in most traffic-prone areas. Different stakeholders query the data in different time frames with varying quantities of data.

5. Apart from the business perspective, smart parking has social significance. Information sharing between multiple vendors to form an interoperable solution is very effective in densely populated areas. If the parking space is filled in infrastructure, the driver shall be directed to a nearby parking location of other vendors. The interoperable parking system is an emerging strategy.

1.2.2 Peer-to-Peer Ridesharing

Peer-to-peer (P2P) ridesharing is a low-cost solution for commuters without a personal vehicle. Public transport including buses and trains operates at a fixed time interval along a specific route, leading to lesser commuting flexibility. Private taxis offer greater flexibility and reliability, but the trip cost is comparatively higher than the public transit and everyone cannot afford it. In P2P ridesharing, the trip owner broadcasts the commuting route and the accommodation capacity to all peers. Hiring a taxi is a trade-off between reliability to affordability.

The user who likes to travel in the same commuting path of a trip owner can subscribe to ridesharing. The P2P ridesharing system finds a matching trip owner in very short notice. This P2P ridesharing emerged in the USA during the 1990s. Parents dropping their children in school and pooling their cars in rotational shifts are some of the examples of unorganized ridesharing along with known people. Due to the lack of technological solutions such as smartphones, Internet, and GPS, ridesharing didn't reach the sharing between strangers. With the improvement in technological solutions, ridesharing is effectively implemented in many cities. When renting a cab or riding our vehicle, only a few of the seats are used. Ride owners are unaware of the people who commute in the same route. ICT bridges the gap between ride owners and commuters. Ridesharing reduces congestion, fuel consumption, and vehicle miles traveled (VMT). Ridesharing can be majorly classified into two types.

Operator-based ridesharing is where the transport network companies (TNCs) such as Uber and Lyft act as a middleman and schedule the ride for a commuter to hop in. P2P ridesharing is in its early stage. To the best of our knowledge, only a few of the platforms are actively using P2P ridesharing. Carma, formerly known as Avego, started as a pilot project in 2007 in Norway. Later in 2011, the project was funded by Transnova. This project aims at corporate ridesharing, verified carpools, and shared mobility services. Quick Ride in India and HentMeg in Norway are actively providing P2P ridesharing services. Based on subscribers, the trip is majorly classified into four types [11].

Type 1: If the starting and ending points of the subscriber are the same as the ride owner, then it is called identical ride sharing.

Type 2: The subscriber hops in the middle of the route and drops in the middle or end of the route similar to the trip owner. In this case, the subscriber's commuting need is satisfied by a single ride owner.

Type 3: The subscriber travel route is covered by multiple ride owners in synchronization. The subscriber drops and opts for one more ride owner since the source and the destination are not covered by a single ride owner.

Type 4: The ride owner takes an alternate route to pick up the subscriber.

In P2P ridesharing, more of the effort is focused on designing efficient algorithms rather than implementation issues [12–14]. The aim is to accommodate more subscribers utilizing fewer vehicle miles traveled (VMT). Many algorithms proposed in the literature also focused on profit maximization. The minimization or maximization problem is formulated as an optimization problem and solved using heuristic algorithms. The literature focused on various objectives that include minimizing the number of transfers, minimizing the waiting time, minimizing the number of ride owners, and increasing the fuel efficiency. Researchers have also done considerable work on combining ridesharing with public transport, where the commuter takes multiple hops including taxis and public transport [15].

Research led by an MIT team has devised an algorithm to match 3,000 passengers to four-seater cars with an average wait time of 2.7 minutes only. The proposed algorithm considers the trade-offs between waiting time, car capacity, fleet size, travel delay, and operational cost. The challenge here they addressed is using a real-time navigation system to compute the travel time between two points to minimize the wait time of another subscriber. The open issues are the fleet rebalancing. The subscriber requests are a dynamic entity.

Challenges and Open Issues

1. Consider a subscriber S1 wants to commute from point A to point B, and a trip owner T1 has only one vacant seat. The subscriber request is mapped to trip owner T1. After some time, one more subscriber S2 wants to travel from an in-between point till the end of the route as the trip owner T1. S2's trip distance is more compared to S1's trip distance. From an economical point of view, serving S2 gives greater profits than serving S1. Instead of T1, S1 is substituted with another trip owner T2 having the same wait time as T1. To achieve this dynamism, many factors need to be considered while designing a heuristic algorithm. The live traffic, driver reliability, surge pricing, and customer reachability are some of the major factors to be considered while designing a dynamic algorithm [16].

2. Ridesharing opens a new door on an individual's security and privacy issues [17,18]. The information gap cannot be bridged by any computer system. To enroll a driver into their platform, Uber performs three-stage background checks on the local criminal database, federal, and state databases. Not all countries are equipped with a centralized system to check the driver's profiles. Sharing a ride

with strangers exposes an individual's personal information, and picking up and dropping points. The entire security concern is abstract in a single question "Is it safe to travel with fellow passengers?" Existing technology can't answer this question due to information unavailability. In recent years, the usage of personal information such as mobile numbers is technologically abstracted. The drivers and passengers are connected only through telephonic exchanges. The driver calls the fixed telephonic exchange, and the call is transferred to the customer without revealing the personal information.

3. Measuring the effectiveness of the ride trip is still a debatable metric in the P2P ridesharing arena [19]. At present, the effectiveness of a ride is measured through user ratings. The unavailability of drivers, sudden changes in passenger plans, rerouted traffic, and unexpected happenings have a great impact on measuring the ride trip. "How reliable the driver is?" The question is answered only through user ratings. We need a reliable procedure to measure the effectiveness of a riding trip.

1.2.3 Self-Driving Cars

Driving is an essential part of everybody's life. Driving needs at the most concentration on roads, or else it leads to collisions. Driving a vehicle over long distances makes the drivers tired. Most of the accidents happen at midnight as the driver falls asleep. As the distance increases, the concentration of the driver decreases. Increasing speed, violating road traffic rules, and lethargic overtaking also have adverse effects on road safety measures. A survey done by the US Department of Transportation in the year 2005 shows that 40,000 people lost their lives due to accidents [20].

The self-driving car is a driverless vehicle that has a real-time object recognition engine and state-of–the-art navigation capabilities. Self-driving cars drive without human interventions following safety rules to a greater extent. In recent years, through artificial intelligence and machine learning (ML) algorithms, the car navigated on the most complicated traffic routes effectively. The idea of the self-driving car emerged in the year 2003, through a competition organized by the Defense Advanced Research Projects Agency (DARPA) for a prize money of $2 million. Initially, the competition is organized to design an autonomous vehicle for use in military and mission-critical applications. In the year 2005, DARPA received 197 prototypes.

The autonomous vehicle needs to travel a difficult desert path of 132 miles. The Stanley team from Stanford won the prize and completed the course in 6 hours 53 minutes traveling at an average speed of 20 miles/hour. The specification of the course is stored in Route Definition Data Format (RDDF) and shared with all the participants. The RDDF file contains road width, speed limits, latitude, and longitude information. GPS, laser for near obstacle identification, color camera for road identification, and RADAR for long-range obstacle identification are used.

The challenge faced in most of the autonomous vehicles is obstacle detection [21,22]. The obstacle ranges from small stones, big rocks, tree trunks, or even potholes. The obstacle detection procedure needs to be carried out in real time. The detection works based on 3D geometry. As the distance between the obstacles and the vehicle decreases, the probability of running through an obstacle is identified with higher certainty. The signals acquired by the sensors are processed for effective decision making. For example, the laser implemented to identify the obstacle should not stop the vehicle that is traveling in the steepest path.

Overhanging trees and tree trunks should also be identified for the safest commuting of an autonomous vehicle.

The next challenges lie in integrating the readings in real time from multiple sensors and positioning systems for effective navigation [23]. In addition to the sensor readings, feeding the location roadmaps and traffic updates helps the navigation even further. Sensors may help the vehicle in localized navigation. The maps and telemeter system help the vehicle take globalized decisions such as choosing a route to reach the destination in a short period or choosing a route with less traffic.

Moving object detection [24] is an unnoticed implementation, but it is the heart of autonomous vehicles. In the object detection phase, the static objects are identified using sensors. The moving object detection is relative to the position of an autonomous vehicle. Consider a man-driven vehicle followed by an autonomous vehicle. Both are traveling at the same speed, without any obstacle in between them. The autonomous vehicle concludes that it's not moving, and it increases the pace, which may lead to a collision. Moving autonomous vehicles around traffic-prone areas need to coordinate with the movement of other vehicles. The roads may have moving objects and static obstacles. The coordinated effort to identify static and moving objects is essential.

The next big challenge faced by autonomous vehicles is lane-keeping [25]. Considering the lanes are marked in clear white and yellow colors, what is the probability of the vehicle not rolling over the curb or the divider? A color camera implemented on the hood of the vehicle continuously sends the image streams. The image streams are processed to identify the width of the road and the size of the lane. The coordinated processing of the positioning system, object detection, and localization system helps the vehicle move in the expected path.

Challenges and Open Issues

1. The computer-aided autonomous vehicle is not equivalent to human drivers. Humans are well adapted to dynamic nature and decide in a matter of seconds, whereas autonomous vehicles need rules and instructions to make decisions. Not all the scenarios can be captured and fed as instructions. The system needs to think and decide in an unexperienced scenario, which is a far long way to achieve [26].

2. The rise of artificial intelligence and ML algorithms helps to achieve the navigation of autonomous vehicles to an extent. In aircraft, the autopilot mode in midair is considered safe because it completely works on deterministic algorithms. The rules and set of instructions are framed as a program. ML algorithms use statistical and probabilistic models to decide. Considering the software system, the Murphy law states that "Anything that can go wrong will go wrong".

3. The ecosystem of self-driving cars is too large and is prone to security attacks [27,28]. This deals with the safety of human lives, an attacker may compromise the system to lead an individual to accidents. During peak hours, if the autonomous vehicles are compromised to crash at prime locations, half of the city roads become inaccessible. In recent years, self-driving vehicles have been fooled not to respond to signboards in the pathways by pasting some stickers in the signboard. The ML algorithm learns from the previously trained model and expects the signboard almost the same as the training set. The vehicle cannot recognize the signboard with mild changes. An object identification experiment was performed to identify the tanks in the war field. The scientist trains the ML model with all the possible

images of tanks from various angles. The model performed well in the laboratory. During the actual testing with foreign datasets, the model can't find the tanks in low lights. Because the model is trained with the photographs acquired in daylight. It is impossible to construct a training model considering all the scenarios in advance with large collective datasets.

4. The equipment implemented in autonomous vehicles is very expensive. It is prone to failure if they are not maintained properly. Ensuring the correct working of the variety of sensor devices is crucial [29]. At present, the implementation is in the inception phase, focused on driving the autonomous car in the right direction. As the techniques mature, the devices are optimized and replaced with compact versions.

5. The unimaginable explosion of digital data and their associated processing. A twin-engine Boeing flight produces around 6 TB of data per hour. The data produced during an entire trip range from 200 TB per flight. The small-scale version of the Boeing flight is the automated driving cars. The vehicle produces more amount of data that cannot be stored and processed in a traditional computing system [30]. Designing of massively parallel processing systems and algorithms is needed. As time passes, the data may become absolute. For example, the latest model flight has a flaw in the engine that it may fail if it continues the consecutive trip. The time frame is very less for processing the massive data. The bandwidth is also a key issue in migrating the collected data to cloud servers for processing.

1.3 IoT and Smart Cities

Smart cities use IoT-based sensors and associated technologies to increase the efficiency and experience of a city. Smart cities ensure greater safety, provide a disease-free environment, and provide clean air and water to the citizens. The data collected from the individuals are correlated to derive an actionable decision. The data uncover the individual's interaction with the city environment. The success of a smart city implementation depends on the bind between the privacy-preserving government policies, supporting technologies, and solution providers.

1.3.1 Telemedicine

The recent advancements in sensor technologies and communication protocols and the development of novel methods to measure human vital signs gave rise to an emerging technology called telemedicine. Using telemedicine, a gradual improvement in mortality rates and reduction in treatment costs are observed in various aspects of treatment cycles [31]. Telemedicine is the key to leveraging effective treatment with limited economic resources. Government organizations are encouraging health facilities to adopt ICT-based methodologies to treat and monitor patients. The health records collected from the patients can uncover a meaningful insight into population-based health statistics. For example, a group of people living near a chemical plant is exposed to a certain gas that causes breathing problems in the long run. If the data are stored as electronic health records (EHRs), it is easy to identify the people affected by ailments due to environmental factors. EHRs help the analyst to isolate the issues with greater effectiveness.

The data that are stored as health records are collected using end-user IoT devices such as low-cost health monitoring systems, activity trackers, and smartphones. EHRs are also appended with the records created using specialized instruments (such as MRI, CT scans, X-rays, and ultrasounds) in health facilities. EHRs are a sharable entity designed to work with different application front ends supporting interoperability. The major content of an EHR is biomedical images and biomedical signals. The biomedical images can be analyzed using ML techniques to isolate the cause in real time. Medical practitioners and doctors are insufficient to cater to the healthcare demands of the world's population. Some of the rarest doctoral specialties are podiatrists (foot and ankle specialists).

Many of the population don't have access to podiatrists because they are very few in number. The data in the form of images and reading collected over a while with countering treatment are fed into a ML training model for treatment suggestions. This technology may not completely replace the job of podiatrists, but it helps a general physician to recommend a counteraction with greater confidence.

The electrocardiogram and electroneurogram are continuous signals that are also very important monitoring metrics. Using IoT-based devices, the effects of drug response on an individual are observed in real time. The drug response history is very helpful in treating other patients with the same medical conditions. Microelectromechanical systems (MEMS), complementary metal-oxide semiconductors (CMOS), and charge coupled devices (CCDs) gave rise to microscale invasive and non-invasive sensors.

For visually impaired individuals, the cameras and hearing aids are incorporated into spectacles for easier navigation. The wireless short-range network enables the spectacle to communicate with ad hoc infrastructure for object detection. A combination of a motion sensor, gyroscope, and sound detectors is used in fall detection of elderly people. The data from individual sensors are correlated to alert the caretakers in real time. Elderly people tend to forget their medication at regular intervals. Using the sensor's readings and contextual information, it is possible to remind them about the medication.

Infectious diseases such as COVID-19, Ebola, dengue, and malaria affect a massive population. In such situations, IoT-enabled devices help us to estimate the spread of diseases. The activity trackers detect the heart rate, the number of steps walked, bodily temperature, and the quality of sleep. The smartphone detects the location of an individual present at an instance. These data are correlated together to perform contact tracing to isolate affected individuals before they spread to others. A recent Google Android update includes a COVID-19 exposure notification using Bluetooth signal strength. An individual smartphone collects Bluetooth MAC addresses anonymously when a person comes into proximity.

If an individual is affected with an infectious disease, the anonymous MAC address individuals are alerted about the exposure. IoT greatly helps in digitizing patient records for efficient storage and retrieval process. Since the data are digitized, it enables multiple health centers to contribute to an individual's EMR. IoT and digitization help to shift from reactive to a proactive treatment strategy. The collected data help in the creation of a clinical decision support system (CDSS). The CDSS helps in optimizing the treatment cycles, which leads to a reduction in time and economic factors.

Challenges and Open Issues

1. Data collected in EMR are heterogeneous in nature. EMR may contain structured, semi-structured, and unstructured data that lead to data integration issues. Correlating multiple sensor readings leads to effective data analytics strategies.

Discontinued time series of data collection is a challenge. For example, an individual used an activity tracker in the past and discontinued it at present. The discontinued data series by itself can't exhibit any insights. Ontology-based contextual correlation of data to fill the missing series of an individual with greater accuracy is the need of the hour [32].

2. Data collected from the individual can be used as a marketing instrument if the security and privacy policies are violated [33]. In this digital era, individuals will be the prime target. When a medical condition of an individual is exposed, and a targeted advertisement is delivered to his inbox, it reduces an individual's confidence levels. Discrimination based on health conditions leads to mental stress and suicide. Even though the data are well encrypted in the computer system, the person who has access to the data may expose it for financial benefits. Integrated digital policies and anonymized record reading are needed to ensure data safety in telemedicine.

3. Improvements are needed for genomic sequence analysis to identify genetically transmitted diseases. Genomic sequencing is a process of searching for an arrangement of four chemical components called bases. The sequence shows possible health complications the individual may suffer shortly. The DNA is extracted, and a coding program is used to convert the structure into a digital form. The digital structure is copied and processed in parallel CPU cores to find matching segments from the library. Sequencing is greatly helpful in the early identification of various cancer strains. Designing an efficient matchmaking algorithm is always a classical problem to solve [34].

4. A large volume of data collected from numerous patients using IoT-based devices cannot be processed in traditional hardware. The data are stored in a cloud-based system to overcome the computational barrier. Designing a resource-optimized ML prediction model to forecast health issues with utmost certainty is needed [35]. The model can also be extended for forecasting the health issues of domestic animals.

5. More than solving the technological aspects of telemedicine, the problem lies in technological adoption, legalities, regulation, and financial transactions in telemedicine. A major part of the rural population lacks the technical knowledge to operate with telemedicine-based devices. Educating and engaging the rural population is also part of IoT-based smart city transition [36,37].

1.3.2 Smart Security

With the emergence of computer vision and ML algorithms in combination with IP cameras, it is made possible to detect a person's identity including his facial expressions. In prime locations such as airports and bus stops, such systems are implemented to detect suspicious activities such as detecting a fast movement against a person or finding an involuntary movement based on compulsion. Such a real-time system is very helpful in detecting and reducing crime activities. The user movement patterns and interactions are recorded to train the ML algorithms.

The ML detects suspicious activities in terms of probabilistic occurrences. Apart from human tracking, IoT sensors are widely used in automated security monitoring in households and commercial buildings. The sophisticated PIR motion sensors, infrared sensors, and IP cameras aid in performing this activity. IoT-based security systems are implemented in two ways. The first is a fully automated security system where nil human interaction is required to identify a security threat. These kinds of implementations are very effective in controlled environments such as banks and offices. After office hours, the security system is activated to monitor the environment. This system raises an automatic alarm if any changes (such as movement, or temperature change) are detected in the environment.

The system automatically sends the suspicious information to law-and-order officials for further action. The second type is a semi-automated IoT-based security system very effective to implement in an uncontrolled environment. The individual interacting with the environment cannot be controlled. For example, in a household implementing a fully automated security system, a false alarm is raised when a pet interacts with the environment. The automated system detects a pet as an intruder and raises an alarm. Instead, when a movement is detected, a notification is forwarded to the user for confirmation. The decision is left to the user to raise an alarm or suppress it.

The semi-automated security system also performs well in public places such as airports. The system alerts the security officials about suspicious activity. The decisions are once again left to the security officials for confirmation. The challenge lies in improving the semi-automated security system with utmost certainty. For example, a movement detection system is implemented in a facility. The system has two possible outcomes: Raise an alarm when an event occurs, or monitor the environment until an event occurs to raise an alarm. When there is no intruder, and the system by mistake raises an alarm, it is termed a false positive or type I error. When the facility is under heist and the system fails to raise an alarm is termed a false negative or type II error. If the type I error increases, the reliability of the system is weakened. The type II error leads to adverse effects.

Challenges and Open Issues

1. Video-based surveillance systems are used in facial detection [38], moving object detection [39], etc. Remote video surveillance gets complicated considering the processing and communication bottlenecks. As suggested in the literature, the best practice to reduce the data is at the edge layers. But, in the video surveillance system, the edge devices can't process the video streams. At the same time, implementing a small-scale processing unit along with the devices increases complexity and is very expensive. Processing the video streams in the data center increases the communication cost. A trade-off-based approach is needed to optimize the cost of implementing edge video processing locations and communication costs.

2. The video-based surveillance systems are mostly utilizing directly powered video cameras. There are some implementations in combination with a motion sensor to stream the events only when it is detected by the sensors. A wireless video-based surveillance system is still unachievable and has wider applications in various fields [40].

3. The use of video compression [41] techniques such as High Efficiency Video Coding (HEVC) improves bandwidth utilization to an extent [42]. HEVC offers high-quality video transmission in lower-bandwidth networks. Such an improvement in video compression techniques is always encouraged.

4. Searching for a specific event in video streams is the hardest part of the surveillance task [43]. Consider an area fixed with several video cameras, and an intelligent agency requires a report on a particular car passed by an area at a specific
time. Manually checking the video streams is time-consuming. Indexing the
videos based on an event occurrence is also a challenge in ensuring security.
Sometimes the stored videos have the required pieces of evidence, but retrieving
the information is time-consuming.

1.3.3 Real-Time Environment Monitoring System

Pollution is introducing harmful contamination by natural means or human activity into
the land, water, and air medium. The land is polluted by nuclear accidents, nuclear weapon
testing, garbage dumping, industrial waste disposal, mining, and chemical manures and
spills. The Fukushima incident in Japan (March 2011), the Chernobyl incident (April 1986),
and the Three Mile Island incident (March 1978) are some of the historically well-known
nuclear accidents. The accidents happen due to human negligence or natural disaster. The
accident has adverse effects on both land and air due to ionizing radiation. This ionizing
radiation has the tendency to change the atoms of living beings from the DNA level, which
leads to cancer.

The improper handling of radioactive waste from nuclear power plants is also a hazard
to the environment. Most IoT solutions are developed for remote monitoring of nuclear
reactors. The most advanced monitoring includes embedded steel corrosion, chloride concentration, and pH levels. Predicting future natural disasters helps us to stall the working
of a nuclear reactor in advance to ensure public safety. Apart from the mentioned, IoT
systems are efficiently integrated into the solid waste collection in a smart city framework.
The dump's capacity is monitored using the sensor. The sensor readings inform the base
station about the filled capacity of the bins. A vehicle route is optimally scheduled to collect the waste from filled dumps.

The location of the dump is integrated with the GPS and the city's roadmaps for scheduling the shortest path to collect garbage waste [the Malmoe simulation model]. Using this
integrated framework, the garbage is evacuated in time based on sensor readings and a
reduction in manpower and collection routes is majorly observed. This also helps in identifying the area where the garbage is filled rapidly to make capacity-based decisions. A GIS
is used to find a landfill where the garbage can be dumped without affecting the public
health. Poison-resistant catalytic pellistor is installed in residential complexes to find gas
emissions beyond prefixed threshold levels. The calibration and placement of such sensors
are very crucial in such high-risk environments. A data logger connects to the pellistor
and collects the data for proactive event prediction.

Nowadays, in developing countries, industrial waste is mixed in water bodies without
pre-treatment. This increases water turbidity, pH levels, and levels of metals and minerals.
Aquatic creatures, domestic animals, and plants are greatly affected due to water pollution. Implementing a sensor on water bodies at a specified distance effectively measures
the water metrics and reports to base stations. A sudden decline in water quality is identified based on sensor geotagging locations. The area is identified to isolate industry that
violates the regulations. Invasive sensors are implemented in drinking water pipelines
to measure the corrosion of pipelines in India. In foreign countries, the public transport
vehicle is implemented with sensors to sense air quality in wider space. This methodology
reduces the number of sensors used, and the sensor is powered using the battery units in
the vehicle.

Challenges and Open Issues

1. The first and foremost issue in implementing the smart city environment monitoring system is the risk of theft or vandalism to the expensive equipment implemented in remote locations. In some cases, the devices are compromised to send misleading data.

2. Device procurement and maintenance tend to be very high for large-scale implementation. Apart from that, integrating the devices from various vendors with different device specifications imposes an additional burden on protocol translation in networking devices [44]. However, using the same specification devices is not impossible while integrating multiple smaller sites of various domains.

3. Retrofitting the existing monitoring strategies in mission-critical applications is not widely accepted in the case of nuclear power plants. The existing system is very reliable, and the engineers don't find a need to replace the system with modern IoT-based sensors. The existing methodology is impenetrable and delivers what it meant to deliver. The existing technology and IoT-based sensors can coexist, but cannot be completely replaced. The fear of reliability is a major concern considering this situation.

1.4 Precision Agriculture

Agriculture is the backbone of the country's economy. Agriculture majorly depends on natural resources such as water, sunlight, and quality soil to grow various kinds of crops. Extreme weather and limited water supply remain a never-ending challenge in farming. Also, the crops continuously need human attention from sowing and watering to harvesting. Watering the crops is a critical activity of farming. Lesser or excessive water may ruin the crop. Countries such as India contribute around 17% of the world population, and yet only 4% of freshwater resources are actively maintained.

The water resources are distributed unevenly across geographical locations. The state of Andhra Pradesh has surplus water, whereas Tamil Nadu depends on monsoon rains and ground water. Ground water is a major source in major cities in and around India, and it is depleting at an alarming rate. Water conservation is practiced all over the country starting from rainwater harvesting, drip irrigation, and sprinkler irrigation. The utilization needs to be optimized as much as possible to meet the future demand. The IoT-based sensor technologies have come to the rescue.

The sensors measure the moisture and temperature levels of the soil. Based on the crop type, a threshold level is fixed in reference to the soil information service database. When the value falls below the threshold, the irrigation system is activated automatically with minimal human interference in the field. The water pumps are made to work with both renewable energies using solar panels and direct electricity. Sprinkler-based irrigation is used in some situations. When the wind is strong, the droplets may miss watering the crops. To overcome this, a system is designed to counteract the wind speed by directing the sprinkler in a certain direction with optimal pumping force.

In some cases, instead of measuring the moisture content, thermal imaging and IR imaging are used to find water shortage in the fields. The matrix-like pipe link structure is activated to irrigate a particular square area. It also has a special advantage of identifying pests in crops. The administration of pesticides is also automated through drones in many developed nations. The automatic pesticide system using drones has the major advantage of covering a vast area in a lesser time frame that arrests the pest spread. Identifying plant diseases at early stages is not possible with human supervision.

The identification of plant disease is possible only when there is a visual appearance at a later point of the plant life cycle. Root diseases such as bacterial and fungi development are even serious in nature and have a direct impact on unprofitable crop growth. Root diseases contribute up to 30%–50% reduction in farm yield. Spectral image sensing can help in identifying the infestation in the early stages even at root levels. The smart agriculture aims not only at the growth and quality of the plants, but also at reducing the number of laborers working in the fields. Indoor farming or controlled environment farming is also on the rise in countries such as Dubai, Singapore, and Malaysia.

Challenges and Open Issues

1. The growth of any plant depends majorly on the soil, water availability, temperature, topology, and pest resistance. For example, the banana tree roots can withstand waterlogged soil, but the plants such as coffee and tea cannot. The tea plant needs moist soil without excessive water; hence, it is mostly planted on hill slopes. The root plants such as carrots and potatoes need a different environment altogether. Identifying a suitable crop considering all the parameters is challenging and needs expert advice. IoT devices integrated with artificial intelligence-based algorithms may help farmers in crop selection procedures [45].

2. Implementing sensors and gateways in a vast agriculture field incurs more cost. The sensors need to be placed in appropriate target locations to minimize the cost constraints. Bio-inspired optimization algorithms are extensively used to minimize the number of sensors used to cover the maximum target area. Notable research is carried out to reduce communication costs and power utilization [46].

3. The digital references available are not enough. Crops and their related factors are based on geographical information including the climate, type of soil, altitude, and topology. Even though oats grow well in India, it's not planted in mass due to less demand. A regional reference database is needed to suggest planting a crop to get maximum benefits including producer and consumer demands. The regional references are subsets of global references. The global references are protocols applicable to all inherited regional references [47].

1.5 Conclusions

In this chapter, we have presented the various applications of IoT and its potential impacts on human life. We have also presented the challenges that are open issues for budding researchers to focus upon. In this article, we have covered transportation, smart cities, and agriculture, but still there are many other fields that IoT has its footprint. In the near future, we would like to present the next part of this article covering the other frontline application fields such as retail, logistics, and smart grid. The security and privacy issues persist in every layer of IoT starting from physical device security in sensor layers to data security in the analytics layer. Regulatory and implementation issues are specific to the regional boundaries.

The next biggest problem the humankind is about to face is unemployment. IoT has made an environment very efficient with a fewer labor force. In future, it is predicted that technology will replace 30% of human laborers in various fields. History repeats once again. During the industrial revolution, the manual human workforce was replaced by machinery. Unemployment has become a greater issue in society. Of course, machines are more efficient than humans and can work 24/7 when provided with electricity or oil. The societal imbalance has already started during this industrial revolution. Wealth started to accumulate the individuals, and laborers were given a minimum wage and later replaced with machinery. Everyone cannot be an intellectual, but everyone needs to be fed. Let us not leave anyone behind in the technological revolution. Technology is to aid humans and not to replace them.

References

[1] Bhaskar, K. B., Prasanth, A., Saranya, P. (2022). An energy-efficient blockchain approach for secure communication in IoT-enabled electric vehicles. *International Journal of Communication System*, 15(11) 1–25.

[2] Kumar, D. R., Krishna, T. A., Wahi, A. (2018). Health monitoring framework for in time recognition of pulmonary embolism using internet of things. *Journal of Computational and Theoretical Nanoscience*, 15(5), 1598–1602.

[3] Prasanth, A., Pavalarajan, S. (2019). Zone-based sink mobility in wireless sensor networks. *Sensor Review*, 39, 874–880.

[4] Jiang, L., Da Xu, L., Cai, H., Jiang, Z., Bu, F., Xu, B. (2014). An IoT-oriented data storage framework in cloud computing platform. *IEEE Transactions on Industrial Informatics*, 10(2), 1443–1451.

[5] Idris, M. I., Leng, Y. Y., Tamil, E. M., Noor, N. M., Razak, Z. (2009). Car park system: A review of smart parking system and its technology. *Information Technology Journal*, 8(2), 101–113.

[6] Idris, M. Y., Tamil, E. M., Razzak, Z., Noor, N. M. (2009). Smart parking system using image processing techniques. *Journal of Information Technology*, 114–127.

[7] Abdulkader, O., Bamhdi, A. M., Thayananthan, V., Jambi, K., Alrasheedi, M. (2018). A novel and secure smart parking management system (SPMS) based on integration of WSN, RFID, and IoT. In *2018 15th Learning and Technology Conference (L&T)* (pp. 102–106). IEEE.

[8] Pham, T. N., Tsai, M. F., Nguyen, D. B., Dow, C. R., Deng, D. J. (2015). A cloud-based smart-parking system based on internet-of-things technologies. *IEEE Access*, 3, 1581–1591.

[9] Yan, G., Yang, W., Rawat, D. B., Olariu, S. (2011). SmartParking: A secure and intelligent parking system. *IEEE Intelligent Transportation Systems Magazine*, 3(1), 18–30.

[10] Prasanth, A., Jayachitra, S. (2020). A novel multi-objective optimization strategy for enhancing quality of service in IoT-enabled WSN applications, *Peer-to-Peer Networking and Applications*, 13, 1905–1920.

[11] Furuhata, M., Dessouky, M., Ordóñez, F., Brunet, M. E., Wang, X., Koenig, S. (2013). Ridesharing: The state-of-the-art and future directions. *Transportation Research Part B: Methodological*, 57, 28–46.

[12] Bei, X., Zhang, S. (2018). Algorithms for trip-vehicle assignment in ride-sharing. In *Thirty-Second AAAI Conference on Artificial Intelligence*.

[13] Tian, C., Huang, Y., Liu, Z., Bastani, F., Jin, R. (2013). Noah: A dynamic ridesharing system. In *Proceedings of the 2013 ACM SIGMOD International Conference on Management of Data* (pp. 985–988).

[14] Huang, Y., Jin, R., Bastani, F., Wang, X. S. (2014). Large scale real-time ridesharing with service guarantee on road networks. *Proceedings of the VLDB Endowment*, 7(14), 2017–2028. arXiv preprint arXiv:1302.6666.

[15] Stiglic, M., Agatz, N., Savelsbergh, M., Gradisar, M. (2018). Enhancing urban mobility: Integrating ride-sharing and public transit. *Computers & Operations Research*, 90, 12–21.

[16] Agatz, N., Erera, A., Savelsbergh, M., Wang, X. (2012). Optimization for dynamic ride-sharing: A review. *European Journal of Operational Research*, 223(2), 295–303.

[17] Gandhi, A., Sucahyo, Y. G., Ruldeviyani, Y. (2018). Investigating the protection of customers' personal data in the ridesharing applications: A desk research in Indonesia. In *2018 15th International Conference on Electrical Engineering/Electronics, Computer, Telecommunications and Information Technology (ECTI-CON)* (pp. 118–121). IEEE.

[18] Aïvodji, U. M., Gambs, S., Huguet, M. J., Killijian, M. O. (2016). Meeting points in ridesharing: A privacy-preserving approach. *Transportation Research Part C: Emerging Technologies*, 72, 239–253.

[19] Abd Elmeguid, S. M., Ragheb, M. A., Tantawi, P. I., Elsamadicy, A. M. (2018). Customer satisfaction in sharing economy the case of ridesharing service in Alexandria, Egypt. *The Business & Management Review*, 9(4), 373–382.

[20] Howard, D., Dai, D. (2014). Public perceptions of self-driving cars: The case of Berkeley, California. In Pedersen, N., Houston, R. (eds) *Transportation Research Board 93rd Annual Meeting*, The National Academies of Sciences, Engineering, and Medicine, Washington, DC, Vol. 14, No. 4502, pp. 1–16.

[21] Uijlings, J. R., Van De Sande, K. E., Gevers, T., Smeulders, A. W. (2013). Selective search for object recognition. *International Journal of Computer Vision*, 104(2), 154–171.

[22] Chen, X., Ma, H., Wan, J., Li, B., Xia, T. (2017). Multi-view 3D object detection network for autonomous driving. In *Proceedings of the IEEE Conference on Computer Vision and Pattern Recognition* (pp. 1907–1915).

[23] Shadrin, S. S., Ivanov, A. M., Karpukhin, K. E. (2016). Using data from multiplex networks on vehicles in road tests, in intelligent transportation systems, and in self-driving cars. *Russian Engineering Research*, 36(10), 811–814.

[24] Cho, H., Seo, Y. W., Kumar, B. V., Rajkumar, R. R. (2014). A multi-sensor fusion system for moving object detection and tracking in urban driving environments. In *2014 IEEE International Conference on Robotics and Automation (ICRA)* (pp. 1836–1843). IEEE.

[25] Chen, Z., Huang, X. (2017). End-to-end learning for lane keeping of self-driving cars. In *2017 IEEE Intelligent Vehicles Symposium (IV)* (pp. 1856–1860). IEEE.

[26] Maqueda, A. I., Loquercio, A., Gallego, G., García, N., Scaramuzza, D. (2018). Event-based vision meets deep learning on steering prediction for self-driving cars. In *Proceedings of the IEEE Conference on Computer Vision and Pattern Recognition* (pp. 5419–5427).

[27] Amoozadeh, M., Raghuramu, A., Chuah, C. N., Ghosal, D., Zhang, H. M., Rowe, J., Levitt, K. (2015). Security vulnerabilities of connected vehicle streams and their impact on cooperative driving. *IEEE Communications Magazine*, 53(6), 126–132.

[28] Lakshmi, D., Srinivas Reddy, G., Manideep, K. (2021). A comparative study on breast cancer tissues using conventional and modern machine learning models. In Satapathy, S. C., Bhateja, V., Favorskaya, M.N., Adilakshmi, T. (eds) *Smart Computing Techniques and Applications. Smart Innovation, Systems and Technologies*, Springer, Singapore, Vol. 225, 693–699.

[29] Boeglin, J. (2015). The costs of self-driving cars: Reconciling freedom and privacy with tort liability in autonomous vehicle regulation. *Yale Journal of Law & Technology*, 17, 171.

[30] Daniel, A., Subburathinam, K., Paul, A., Rajkumar, N., Rho, S. (2017). Big autonomous vehicular data classifications: Towards procuring intelligence in ITS. *Vehicular Communications*, 9, 306–312.

[31] Zanaboni, P., Wootton, R. (2012). Adoption of telemedicine: From pilot stage to routine delivery. *BMC Medical Informatics and Decision Making*, 12(1), 1–9.

[32] Peral, J., Ferrandez, A., Gil, D., Munoz-Terol, R., Mora, H. (2018). An ontology-oriented architecture for dealing with heterogeneous data applied to telemedicine systems. *IEEE Access*, 6, 41118–41138.

[33] Jin, Z., Chen, Y. (2015). Telemedicine in the cloud era: Prospects and challenges. *IEEE Pervasive Computing*, 14(1), 54–61.

[34] Bankevich, A., Nurk, S., Antipov, D., Gurevich, A. A., Dvorkin, M., Kulikov, A. S. … Pyshkin, A. V. (2012). SPAdes: A new genome assembly algorithm and its applications to single-cell sequencing. *Journal of Computational Biology*, 19(5), 455–477.

[35] Shickel, B., Tighe, P. J., Bihorac, A., Rashidi, P. (2017). Deep EHR: A survey of recent advances in deep learning techniques for electronic health record (EHR) analysis. *IEEE Journal of Biomedical and Health Informatics*, 22(5), 1589–1604.

[36] Weinstein, R. S., Lopez, A. M., Joseph, B. A., Erps, K. A., Holcomb, M., Barker, G. P., Krupinski, E. A. (2014). Telemedicine, telehealth, and mobile health applications that work: Opportunities and barriers. *The American Journal of Medicine*, 127(3), 183–187.

[37] Scott Kruse, C., Karem, P., Shifflett, K., Vegi, L., Ravi, K., Brooks, M. (2018). Evaluating barriers to adopting telemedicine worldwide: A systematic review. *Journal of Telemedicine and Telecare*, 24(1), 4–12.

[38] Grgic, M., Delac, K., Grgic, S. (2011). SCface–Surveillance cameras face database. *Multimedia Tools and Applications*, 51(3), 863–879.

[39] Kalli, S.N.R., Suresh, T., Prasanth, A., Muthumanickam, T. (2021). An effective motion object detection using adaptive background modeling mechanism in video surveillance system. *Journal of Intelligent & Fuzzy Systems*, 41(1), 1777–1789.

[40] Räty, T. D. (2010). Survey on contemporary remote surveillance systems for public safety. *IEEE Transactions on Systems, Man, and Cybernetics, Part C (Applications and Reviews)*, 40(5), 493–515.

[41] Bairagi, V. K. (2017). Big data analytics in telemedicine: A role of medical image compression. In Márquez, F. P. G., Lev, B. (eds) *Big Data Management* (pp. 123–160). Springer, Cham.

[42] Sullivan, G. J., Ohm, J. R., Han, W. J., Wiegand, T. (2012). Overview of the high efficiency video coding (HEVC) standard. *IEEE Transactions on Circuits and Systems for Video Technology*, 22(12), 1649–1668.

[43] Hu, W., Xie, N., Li, L., Zeng, X., Maybank, S. (2011). A survey on visual content-based video indexing and retrieval. *IEEE Transactions on Systems, Man, and Cybernetics, Part C (Applications and Reviews)*, 41(6), 797–819.

[44] Xiao, G., Guo, J., Da Xu, L., Gong, Z. (2014). User interoperability with heterogeneous IoT devices through transformation. *IEEE Transactions on Industrial Informatics*, 10(2), 1486–1496.

[45] Balezentiene, L., Streimikiene, D., Balezentis, T. (2013). Fuzzy decision support methodology for sustainable energy crop selection. *Renewable and Sustainable Energy Reviews*, 17, 83–93.

[46] Zhang, Y., He, S., Chen, J. (2015). Data gathering optimization by dynamic sensing and routing in rechargeable sensor networks. *IEEE/ACM Transactions on Networking*, 24(3), 1632–1646.

[47] Shah, P., Hiremath, D., Chaudhary, S. (2017). Towards development of spark based agricultural information system including geo-spatial data. In *2017 IEEE International Conference on Big Data (Big Data)* (pp. 3476–3481). IEEE.

2

Introduction to Cognitive Computing

M. Nalini
Sri Sairam Engineering College

A. Prasanth
Sri Venkateswara College of Engineering

Arunkumar Gopu
VIT-AP University

D. Lakshmi
VIT Bhopal

CONTENTS

DOI: 10.1201/9781003256243-2

2.1 Introduction

Cognition means acquiring knowledge and understanding it by processing things such as thoughts, senses, and experience [1]. The word cognition comes from the Latin word cognito. This term covers many peculiar functions such as compiling the collected data, understanding, making decisions, analysing the problem, and identifying solutions from experience. On the whole, this term refers to extracting new information from the available collected information. When technology is included in the word cognition, it becomes cognitive computing. It is a technology that tries to work as well as the human brain for processing the collected data and giving appropriate action. This technology is not primarily intended to mimic the human brain; rather, it is intended to provide proper assistance to the human.

The fundamental principle of cognitive systems is their intelligent systems, which rely on and are interconnected with a variety of specialized functions [2]. For example, when a person sees a picture, the respective signal will go to the brain and, based on the content of the picture, the person will react. If it's a bad picture, then the person will try to avoid or close their eyes immediately. This reaction is because of learning experiences from past situations. Cognitive computing tries to bring this human intelligence to a system.

Researchers understand that this intelligence of the human brain is attained from various interrelated processes such as language, perception, emotion, and memory, and they developed this cognitive computing, which includes various processes such as machine learning, deep learning, natural language processing (NLP), IoT, big data, contextual awareness, cloud computing, reasoning, and visualization. These techniques make the system gather data like the sensory organs in a human and can understand, remember, analyse, and make decisions like a human brain. On the whole, the cognitive computing system is depicted in Figure 2.1 with three important roles.

2.2 Evolution of Cognitive Computing

Artificial intelligence (AI) is the basis of cognitive computing. AI has been around for a long time, but in the last 50 years, there has been tremendous growth in the field of cognitive computing. The first smart programs were developed in the 1950s by McCarthy and Marvin Minsky. In the period between the 1960s and 1970s, the foundations of AI were developed by Ray Solomon. He developed Bayesian algorithms for prediction and inductive interference. In the 1970s, the concept of perceptron was developed by Minsky and Papert, which is the basis for neural networks. NLP research was started in the same period by Robinson and Walker.

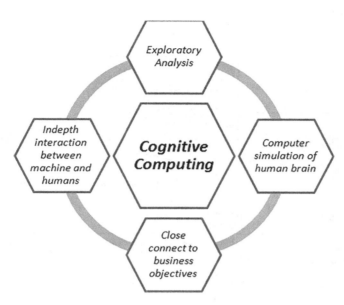

FIGURE 2.1
Cognitive computing.

The first machine learning algorithm for NLP was developed in the 1980s. All the important technologies required for AI were developed during the 1990s period, and in the period of 2000s, almost all of the areas required for cognitive computing boomed. In 2011, IBM Watson, one of the major cognitive computing systems, also started in the period of 2000s, and in 2011, Watson proved its cognitive success in the Jeopardy! game show. Cognitive computing is a very complex field, so much so that it requires many fields of knowledge [3]. Figure 2.2 shows the cognitive computing evolution process stage by stage. Data discovery is an early stage of cognitive computing; more good data discovered means greater system efficiency and accuracy, so data discovery is important in cognitive computing.

After data discovery, the information should be processed like in the human brain. For that, cognitive science is involved, which consists of various functions like the brain. It's an interdisciplinary science where different domains are combined to try to achieve the intelligence of the human brain. The human brain is a multiple-part system to achieve a peculiar mental process, which consists of visual, auditory, memory, perception, imagination, language understanding, problem-solving ability, and attention.

Cognitive science researchers perform experiments under some assumptions and restrictions. From there, they develop algorithms for that particular task. These algorithms were used to develop intelligence in cognitive systems [4]. As we know, data play a major role, so the next technology used in cognitive systems is big data. Big data is very huge and cannot be generated using traditional methods. Big data can be generated by machines, humans, or nature, and we can select and collect the required source based on the application. With recent technology, the data can be produced as organized, semi-organized, and unorganized from various sources. The characteristics of big data are classified as 5Vs as shown in Figure 2.3.

One important thing about big data in cognitive computing systems is that the data need not be big. A small amount of data can also be used based on the application. For example,

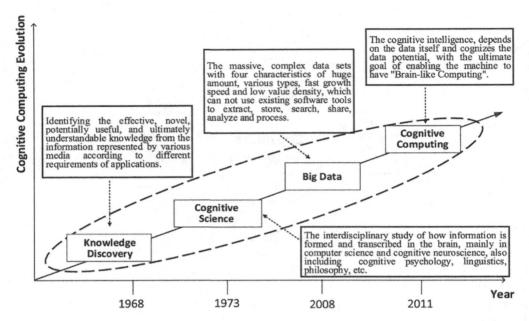

FIGURE 2.2
Cognitive computing evolution.

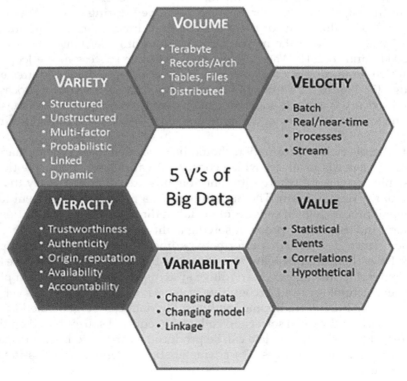

FIGURE 2.3
Features of big data.

if a particular process gives only a limited amount of data, but is in need of cognitive systems, it can also be achieved using some dedicated algorithms. With the help of all this technology, the system can achieve cognitive ability like the human brain. These technologies should be strengthened and need more research to achieve brain-like computing.

Now, to make the system more interactive with human beings, it is connected to the Internet of things (IoT) [5–8]. With the use of sensors, a huge amount of data can be generated in different forms, which is very useful for developing an effective cognitive system. The data collected should not be wasted, so careful data handling techniques are required to process the collected data because most of the data collected today are unstructured and semi-structured data.

2.3 Comparison

Cognitive computing is a subcategory of AI. The technologies are similar for both CC and AI. When comparing CC and AI, it's almost the same, but some small differences are there.

1. Cognitive computing tries to mimic human intelligence, but AI is for solving a particular problem using the best algorithm and provides an accurate result.
2. Cognitive computing is not for making decisions for humans; instead, it is used for giving assistance to humans in decision-making. On the contrary, AI is used for taking decisions; for example, in the healthcare industry, AI is used to identify preferable treatments.

2.3.1 Example Case

Imagine that a person wants to change their present job. In this case, the AI assistance will decide on a matching job based on that person's skills, suitable job, and pay. The AI assistance will take that decision on behalf of the person. Figure 2.4 shows both approaches.

However, cognitive assistance will suggest the suitable jobs available based on the person's career growth, salary expectation, and qualification, now that particular person is responsible for taking decisions based on the suggestions given by cognitive assistance. So, cognitive computing helps to make smarter decisions.

2.4 Dimensions of Cognitive Computing

Cognitive computing consists of both hardware and software depending on the application [9]. This consists of different dimensions, which are

- Reasoning
- Learning
- Deduction

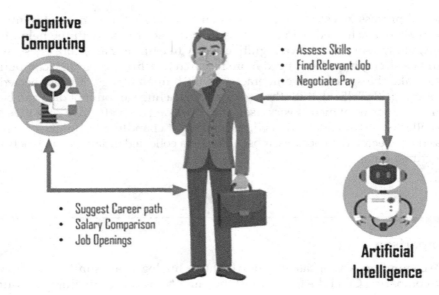

FIGURE 2.4
Example case.

- Perception
- Inference
- Motivation
- Reflection.

From Figure 2.5, we can visualize the dimensions of cognitive computing one by one.

2.4.1 Reasoning

In this dimension, the system should understand and analyse complex problems so that it can identify the possible solution to the given problem. Pattern recognition techniques are one example of reasoning in cognitive computing.

2.4.2 Relating

In this case, the system will identify how to communicate with humans regarding results. Communication with humans can be done using AI techniques such as NLP. Nowadays, numerous techniques are available based on the application and availability. We can choose one.

2.4.3 Perception

As humans, we understand the real world using our sensory organs. The human brain processes the signals from sensory organs and understands the environment. Like us humans, the system will understand the real world using sensors. Sensor outputs are combined as data, which might be structured or unstructured. The collected information will be processed to understand the situation.

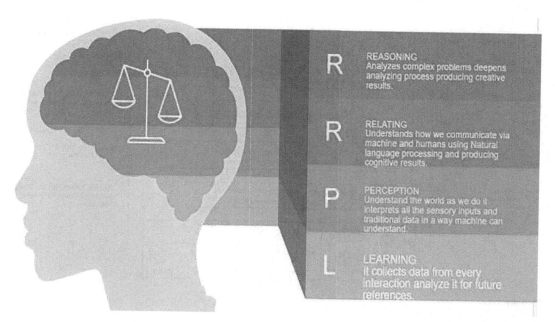

FIGURE 2.5
Cognitive computing dimensions.

2.4.4 Learning

The system collects information from every interaction, and the collected information will be analysed and used for future interactions.

2.5 Architecture of Cognitive Computing

Cognitive computing architecture is shown in Figure 2.6, which shows the technologies involved, system requirements, and technology challenges. In Figure 2.6, the top layers show application-oriented cognitive systems for some specific examples such as healthcare, smart cities, and clever transportation systems. For each and every example, there is a technology challenge and a system requirement that will be discussed in detail.

2.5.1 IoT in Cognitive Computing

As we know, cognitive computing is mainly dependent on data. Transformation of information can be done using communication technology, and utilization of information will be performed using computers. Information is usually called data, which may be in any form, structured or unstructured. IoT connects data with the system, and it eliminates the barriers between systems and humans. The pillars of cognitive IoT (CIoT) are shown in Figure 2.7.

The main objective of CIoT is to improve the performance of the IoT in terms of its intelligence using cognitive computing technology. The stand-alone IoT is mainly for collecting data from its surroundings and giving reactions based on the pre-program. If an IoT

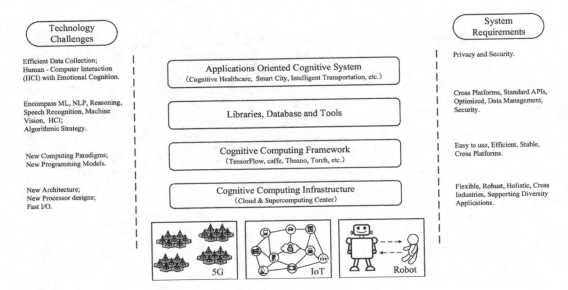

FIGURE 2.6
Architecture of cognitive computing.

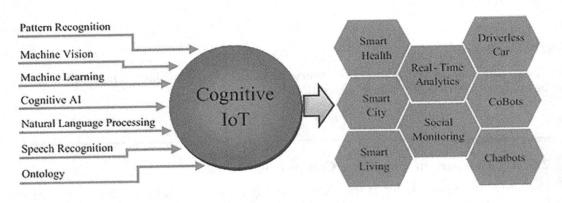

FIGURE 2.7
Cognitive IoT (CIoT).

system faces any situations that are not pre-programmed, then it fails to react. To improve and introduce intelligence to IoT, cognitive computing is added to IoT, then it becomes a fully autonomous system [10].

2.5.2 Big Data Analysis in Cognitive Computing

In this digital era, an enormous quantity of information is produced from various sources. It may be organized, semi-organized, or shapeless data, and constitutes cognitive big data. During the processing of data, it creates a problem. This can be solved by using CC techniques. Big data analysis in association with cognitive computing tries to mimic the big data thinking of the human brain [11].

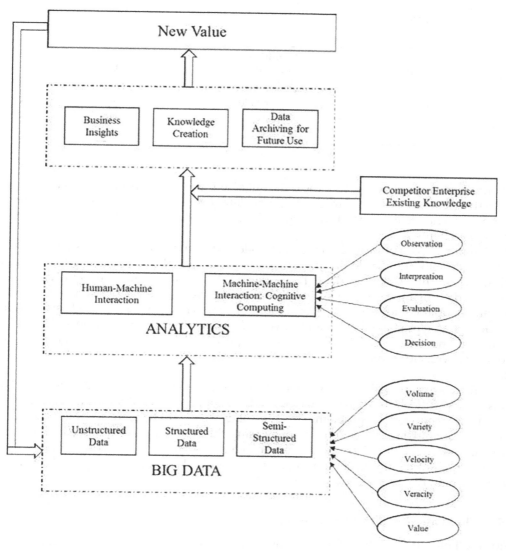

FIGURE 2.8
Conceptual model.

There are three levels of human thinking. The first level is about their physical lives and surroundings, the second about spiritual life, and the last about the value of life. The number of upper-level thinkers is very small. Now, cognitive computing develops machine intelligence only at the first and second levels. Using machine intelligence, human standards of living can be improved. Significant applications related to this are smart healthcare, smart homes, and health monitoring. Attaining third-level intelligence is a challenge for AI.

The conceptual model of big data analytics and CC is shown in Figure 2.8. It shows the relationship between cognitive computing and big data clearly. Data are the mandatory requirement for cognitive computing, so data observation is the first step. Data collection

TABLE 2.1

Characteristic Mapping

S. No.	Cognitive Computing	Big Data
1	Observation	Volume
2	Interpretation	Variety
3	Evaluation	Velocity
4	Decision	Veracity

not only means simply collecting the data, but also involves scrutinizing the correct data. For getting better results, a large volume of data are required. Today's data should be normalized and cleaned to get the required data, and this can be done with the help of cognitive systems.

Different varieties of information can be collected from various sources such as the IoT and social media [12]. Based on the source of data and application, the cognitive system will interpret this variety of data. The collected data will be converted into visual networks of data using cognitive systems. This will provide the relationship between the data and should be done in a shorter amount of time. This process is called "evaluation". The next characteristic of big data is its velocity. Here, the data generation speed will be monitored and, based on this, the processing speed will be decided. The next significant property is veracity, which ensures the genuineness of data in terms of their quality and trust.

Cognitive systems will make decisions based on the analysis done using the data. This decision depends on the evidence based on different predictive algorithms when there is no clear-cut picture of the problem. This type of decision will be more accurate when compared to human trial-and-error decisions. Using this, the mapping sandwiched between the characteristics of big data and CC can be determined, which is shown in Table 2.1.

The growth of cognitive computing along with big data has so many applications in today's human life.

2.5.3 Cloud and Cognitive Computing

With the help of big data technology, huge quantities of information are collected and the collected data will be processed using cognitive computing technology. To do this, a large amount of storage is required. This large amount of storage can be made possible using cloud computing [13]. Using this, the required data can also be retrieved whenever it is required using the Internet. Google Cloud is one example of cloud computing. So, using cloud and cognitive computing, the best suitable solution can be found for the given problem.

On the whole, cognitive cloud computing plays a major role in the current market. The contribution of cognitive computing in various important sectors is shown in Figure 2.9.

Major cognitive computing market companies include Cisco, Microsoft, SAP, Nuance Communications, and IBM. Three categories of cloud services are available, which are software-as-a-service (SaaS), infrastructure-as-a-service (IaaS), and platform-as-a-service (PaaS). Based on the application and requirements, anyone can be used. Different types of clouds are also available, which are private, public, and hybrid clouds. These can also be selected based on the application and cost.

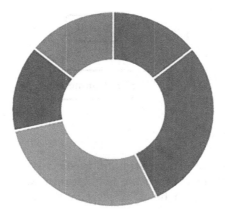

■ Product Development ■ Mergers & Acquisitions ■ Joint Venture

■ Product Launch ■ R&D Investment

FIGURE 2.9
Five forces analysis.

2.6 Supporting Technologies for Cognitive Computing

Deep learning and reinforcement learning are two important supporting technologies for cognitive computing. Deep learning will learn using the high-level features, whereas reinforcement learning will learn from its environment and surroundings, as you will see in this section.

2.6.1 Reinforcement Learning

One type of machine learning technique is supervised learning, and the other one is unsupervised learning. Both the techniques were trained with limited data in a closed form. But in a real situation, the data received may differ when compared to the data used for training. In that case, both the techniques were inefficient to get the solution. So, to improve intelligence, reinforcement learning was introduced. It's a type of machine learning technique that learns and behaves based on feedback. The agent reacts based on the feedback received from the environment. The agent receives a reward for every correct action, and he/she receives a penalty for every wrong action. This is almost similar to our human brain.

For example, when an adult trains a child to learn a new word, if the child says it correctly, the adult will reward the child, but if the child says it incorrectly, the adult will give negative feedback so the child can learn from this feedback. This is called reinforcement learning. Figure 2.10 shows its block diagram. In this example, a child is an agent. The reward system is the environment.

2.6.1.1 Components

1. **States:** the condition of the environment after the action of an agent.
2. **Action:** performed by the agent by observing the environment.
3. **Reward:** the feedback received from the environment based on the action of the agent. The agent can learn from this reward.

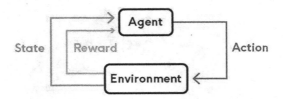

FIGURE 2.10
Block diagram of reinforcement learning.

In this type of learning, both successful and unsuccessful attempts have an impact. This learning process is continued up to the goal being reached or the condition being met. There are three types of approaches to implement this learning problem.

2.6.1.2 Approaches

To solve problems using reinforced learning, there are three approaches.

2.6.1.2.1 Policy-Based

In this approach, the agent behaves based on the policy given to that particular system. The policy must be optimized because the agent performs based on the policy, which is given by

$$a = \pi(s)$$

where

   a – action

   s – state

   π – policy function.

Learning of policy function is based on the best action. There are two types of policies, which are as follows:

   i. Stochastic policy
   In this, the agent will learn based on the rewards it receives for its action. Based on the amount of reward, it learns the quality of the action. It's a probability distribution based on the different actions, and it is represented using the following expression.

$$\pi(s) = \mathbb{P}[a \mid s, \theta]$$

   In the state s, action a will be decided based on the parametric vector value θ; the sample distribution is shown in Figure 2.11.
   ii. Deterministic policy
   The agent will perform the same action for the particular state. It works based on the preset conditions, which are given by the equations as follows:

$$S = (s) \cdot A = (a)$$

   This policy is suitable for a deterministic environment where there is no chance of uncertainty, for example the movements of chess coins in a chess game.

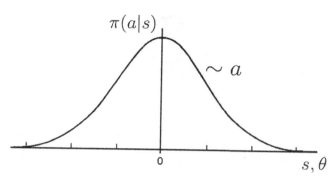

FIGURE 2.11
Distribution of action *a*.

2.6.1.2.2 Value-Based

In this approach, the optimum value for the given environment will be calculated, and based on this, an action will be taken by the agent. Each state's value is the total reward an agent can expect to collect over the future from a particular state.

$$V\pi\ (s) = E\pi\big[R\ (t+1) + \gamma R(t+2) + \gamma\wedge 2R(t+3) + \cdots | St = s\big]$$

Based on the above equation, the agent will choose the action for every step. The agent will always choose the state that has the bigger value.

2.6.1.2.3 Model-Based

In this, the model will be created, which replicates the working environment. So, for different environments, different models will be created. This policy is widely used for developing embedded systems.

2.6.2 Cognitive Computing and Deep Learning

The human brain's cerebral cortex is divided into two hemispheres, which are called the right and left brain. Both have different purposes. In this, the left brain is in charge of logical and rational thinking and the right brain is in charge of creative and visual thinking. The thinking capacity of the brain is divided into two, which are rational and conceptual thinking. Rational-level thinking is based on our beliefs and facts, and the perceptual level of thinking is based on input and output mapping. This thinking ability of humans can be mimicked using data science in association with cognitive computing. An example is shown in Figure 2.13, which depicts both the types of thinking.

In Figure 2.12, the identification of a square is presented. In this section, the square is identified based on its sides, which means it strictly depends on the mathematical definition of square. Section b, on the other hand, identifies the square based on perception. The image of a square is shown to the child many times and taught that it is a square, so whenever the child sees the same, he/she can recognize it's a square. Without knowing the concept of square, now the child can recognize it, but after a number of practices, the child can learn the relationship between all sides and angles of that square, and the child can understand the concept also.

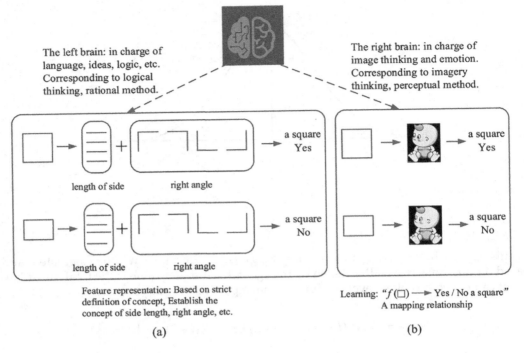

The left brain: in charge of language, ideas, logic, etc. Corresponding to logical thinking, rational method.

The right brain: in charge of image thinking and emotion. Corresponding to imagery thinking, perceptual method.

length of side   right angle

a square
Yes

length of side   right angle

a square
No

a square
Yes

a square
No

Feature representation: Based on strict definition of concept, Establish the concept of side length, right angle, etc.

Learning: "$f(\Box) \longrightarrow$ Yes / No a square" A mapping relationship

(a)   (b)

FIGURE 2.12
Example for rational (a) and conceptual thinking (b).

2.6.3 Cognitive Computing and Image Processing

When the system is used to solve real-world problems, it should behave like a human brain when solving the same. Feature extraction for image identification problems can be solved using either logical thinking or creative thinking [14]. In logical thinking, the features of the image will be extracted using the feature design of the image or using some predictive models. In creative/visual thinking, the extraction of features can be done using deep learning techniques. In today's world, AI is a much needed technology in all fields. Our human brain understands real-world problems very fast, but it is not easy to define the same problem using a rational method for computers. Therefore, defining image features manually is very difficult.

There are two types of feature extraction: manual and automatic. In the manual method, the defining of features in terms of dimensions is very difficult and it takes more time to get a decision. To understand the images quickly, automatic feature extraction techniques were introduced. It uses special algorithms or deep learning networks to extract the features of the image automatically without using human help.

2.7 Cognitive Analytics (Coganalytics)

Cognitive analytics combines both data analytics and visual analytics to make the system fully automatic, and it removes the human from the loop [11]. It has more analytical

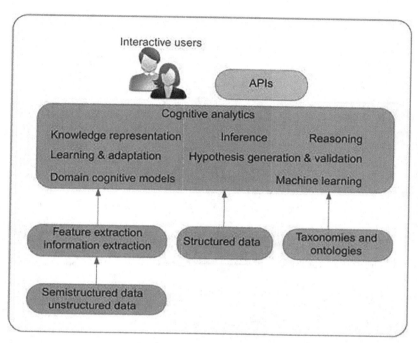

FIGURE 2.13
Internal structure of cognitive analytics.

techniques to analyse large amounts of data. It converts unstructured and semi-structured data into structured data. It deals with both high-level and low-level feature extraction to make the system accurate. A functional block diagram of cognitive analytics is shown in Figure 2.13. It consists of internal mechanisms of the system. It uses various knowledge representation techniques and inferences, which include AI algorithms and semantics, to understand and solve the given situation. It generates multiple answers with its confidence level.

Cognitive analytics performs like a human brain to solve the given task. An architecture is proposed in [15], which is shown in Figure 2.14. Coganalytics is a term used to represent cognitive analytics. This consists of eight layers. The first layer is a physical data layer, which is responsible for data collection and storage. This layer receives data from different sources based on the application, such as WordNet and DBpedia. The data can be static or dynamic; static data can be physically saved, while dynamic data can be conceptually stored. Meanwhile, because the amount of data is so vast, a good data management system is essential.

The physical hardware layer is the second layer, and it is responsible for high-performance computing and processing processes. It improves its infrastructure using neuromorphic devices and neural network accelerators. The third layer is the hardware abstraction layer, which acts as a virtual mechanism on top of the actual hardware layer. The libraries and frameworks allow programmers to write code without having to worry about the underlying special hardware. To ensure efficient execution, the application code is automatically converted. At the moment, Hadoop and Spark are prominent options for implementing this layer.

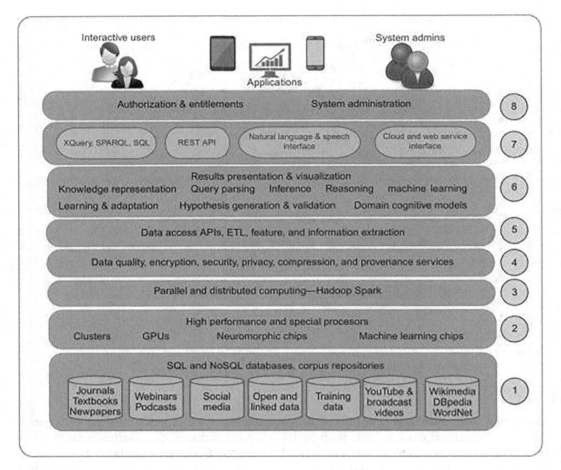

FIGURE 2.14
Architecture of coganalytics.

Level 4 is a low-level data services layer that is in charge of data services such as data cleansing, compression, encryption, and data quality evaluation. It also protects the privacy and origin of the data. Maintaining data privacy is a major challenge. In the healthcare industry, the idea of individually recognizable data is very significant. Data perturbation, for example, allows data analytics to be done without jeopardizing privacy. Data perturbation is an effective strategy for safeguarding the confidentiality of electronic health information over de-identification and re-identification.

The high-level data services layer (Level 5) is where top-level information processing occurs. It includes ETL (extract, transform, and load) tools for high-level data processing. It also includes libraries and software tools for extracting data. KETL, Scriptella, Talend Open Source Data Integrator, Pentaho Data Integrator, Kettle, Jaspersoft ETL, GeoKettle, Jedox, Apatar, CloverETL, and HPCC Systems are some of the well-known ETL systems.

The basic layer of this coganalytics architecture is Layer 6. It includes a variety of machine learning methods, domain cognitive models, inference and perceptive processes, as well as spatial and temporal reasoning mechanisms. Several knowledge representation techniques are presented to aid inference and reasoning.

For parsing queries and detecting subqueries, the Query Parsing subsystem is in charge. The Hypothesis Generation & Validation subsystem is in charge of generating many solutions to a problem and giving a level of assurance to each one. The demonstration and visualization area provided is responsible for presenting the results. It also includes visualization tools for exploring the consequences of features and their functioning. The cognitive analytics layer is the name of this layer.

The API layer is the seventh layer. Declarative query languages and APIs provide access to both interactive users and other systems. Both natural language text and spoken language can be used to specify queries. Coganalytics functionalities are also exposed as cloud and web services through this layer. These services make it possible to create cognitive analytics apps without the need to know anything about programming.

Layer 8 is known as the administrative layer, and it has two main purposes. One is the administrative subsystem, which contains utilities for generating users and assigning responsibilities to them. The Authorization & Entitlement subsystem is the next subsystem in the same layer that is answerable for verifying customers and confirming that they only perform operations for which they have been granted permission.

2.8 Applications of Cognitive Computing

The increasing power of cognitive computing has made so many applications [16]. In future, the growth of cognitive computing is substantial. Industries embed cognitive technology to make their products or services more effective, convenient, safer, faster, and more valuable. In addition to this, it involves creating new products or services and creating a new market. We will explore some important applications here. Figure 2.15 shows the applications of cognitive computing in different areas.

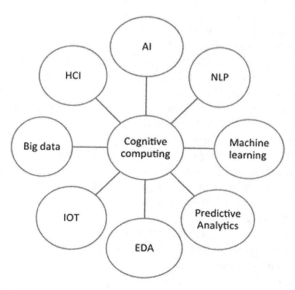

FIGURE 2.15
Applications of cognitive computing.

2.8.1 Cognitive Computing in Robotic Industry

Cognitive RPA refers to RPA tools and solutions that use AI technologies such as optical character recognition (OCR), text mining, and machine learning to increase the experience of users. The robotic system can interpret the client's purpose, make use of the unstructured data related to the consumer, forecast behaviour, and then accomplish a demand in the backend thanks to cognitive RPA capabilities. This can also keep track of the entire client history and thread for further action.

To summarize, cognitive RPA's strength lies in its capability to analyse shapeless data such as papers and emails and then utilize this information to generate increasing degrees of computerization. Some examples are given below:

- Image identification and OCR
- Extracting and identifying targets
- Text analytics
- Emotional investigation
- Labelling
- Grouping
- Speech recognition.

In the future, the next generation of robots will mimic human beings in almost all aspects. In particular, there would be a partnership relationship in which humans and robots live together in coordination and create their corresponding benefits complementary to one another. Co-fusion with humans is a key aspect of the next generation of robots.

2.8.2 Cognitive Computing in Emotion Communication

Sentiment analysis is the discipline of deciphering emotions expressed in written or spoken communication [17]. While humans find it easy to comprehend tone, intent, and other aspects of a conversation, robots find it significantly more difficult. You must give training data of human speech to computers in order for them to comprehend human communication and then assess the correctness of the analysis. Sentiment analysis is commonly used to examine social media communications such as tweets, comments, reviews, and complaints.

The advanced level of picture analysis is face detection. To distinguish a face from another, a cognitive system employs data such as its structure, features, and eye colour. Once a facial picture has been created, it may be used to recognize a person's face in a photograph or video. While this was formerly the case, it is no longer the case.

In particular, cognitive computing is notable for its frequent contact between academics and industry, particularly in education. Companies such as IBM are using cognitive computing to leverage the potential of big data in a variety of fields, including education. In EDM, predicting student success based on a variety of criteria has been a significant study topic (education data mining).

Chatbots are computer programmes that can mimic human conversations by understanding the context of the message. NLP, a machine learning technology, is employed to make this happen. NLP allows computers to receive human input (voice or text), evaluate it, and then respond logically. Cognitive computing allows chatbots to have a certain level of intelligence. Thanks to cognitive computing, chatbots can communicate with a certain amount of intelligence, for example recognizing the user's needs based on previous contact and making ideas.

2.8.3 Cognitive Computing in Retail and Logistics Industry

The retail business offers a lot of fascinating possibilities for cognitive computing. It enables the marketing team to collect more data and analyse it in order to improve retailer efficiency and adaptability [18]. These assist businesses in increasing sales and providing customized recommendations to clients. E-commerce sites have done an excellent job of integrating cognitive computing; they collect some basic information from clients about the fundamental details of the product they are searching for, analyse the enormous amount of data accessible, and then propose goods to the user. Cognitive computing has ushered in a slew of new developments in the sector. Cognitive computing has given retailers the ability to build more flexible organizations through demand forecasting, price optimization, and website design through demand forecasting, price optimization, and website design.

In the transportation, logistics, and supply chain industries, cognition is the new frontier. It is beneficial at all stages of logistics, including warehouse design, management, and automation. In this process, cognition aids in the compilation of packing codes, automatic picking using an autonomous guided vehicle, and the usage of warehouse robots. Logistics distribution networks employ cognition to plan the optimum path, increasing recognition rates and reducing labour costs. The IoT will aid warehouse infrastructure management, inventory optimization, and warehouse operations, and an autonomous guided vehicle will be employed for picking and placing activities. Wearable devices, in addition to IoT, are a significant technology that helps to transform human decision-making in the area of warehouse operations.

2.8.4 Cognitive Computing in Banking and Finance

In the banking business, cognitive computing will aid in improving operational efficiency, client engagement, and experience, as well as increasing revenue [16]. Deeper contextual interaction, new analytics insights, and enterprise transformation are all aspects of cognitive banking that will totally revolutionize banking and financial organizations. We're already seeing signs of this shift in chores such as digitally completing different financial transactions, creating a fresh retail account, and handling claims and loans very quickly.

In the fields of product administration and consumer service assistance, this technology has been shown to be quite beneficial. It will also enable a tailored connection between the financial institution and the client by engaging with each customer on an individual basis and concentrating on their needs. Based on the other materials accessible online, the computer will intelligently identify the customer's personality.

2.8.5 Cognitive Computing in the Power and Energy Sector

The next intelligent future is termed "smart power". The oil and gas sector is under enormous cost pressure to discover, produce, and distribute crude oil and its by-products [19]. They also have a scarcity of qualified engineers and technical personnel. Energy companies must make a variety of key decisions involving large sums of money, such as which site to investigate, how to allocate resources, and how much to produce. This choice has been made for a long time based on the data collected and kept, as well as the project team's skill and intuition.

Technologies that use cognitive computing use volumetric data to aid judgments and learn from the outcomes. This technology will assist us in making key future decisions such as economically viable oil wells and strategies to improve the efficiency of current power plants.

2.8.6 Cognitive Computing in Cybersecurity

Cognitive algorithms develop end-to-end security technologies that identify, analyse, investigate, and mitigate threats [20]. It will aid in the prevention of cyber-attacks (also known as cognitive hacking), making clients less prone to manipulation and providing a technical method to identify any false data or misinformation.

With the rise in volumetric data, cyber-attacks, and a scarcity of trained cybersecurity specialists, new approaches such as cognitive computing are needed to combat these cyber-dangers. Cognitive services for cyberthreat identification and security analytics have already been developed by major security providers in the sector. These cognitive systems not only identify threats, but also evaluate systems, check for weaknesses, and make recommendations.

Alternatively, because cognitive computing needs a huge amount of volumetric data, protecting the data's privacy is critical. To fully benefit from cognitive computing, we must first create a big database of data while simultaneously maintaining its secrecy and preventing data leakage.

2.8.7 Cognitive Computing in Healthcare

Healthcare has ushered in a cognitive computing revolution to aid medical professionals in improving disease treatment and patient outcomes [21–23]. The cognitive computing system analyses massive volumes of data in real time to respond to particular questions and provide intelligent recommendations. Patient involvement has improved as a result of cognitive computing, as has service accessibility.

The effect of success may be calculated by drawing an iron triangle with one side representing efficiency in diagnosis time, one more side representing budget savings, and the third side representing individualized care. The iron triangle's three sides should concentrate on management. The iron triangle's three sides should be focused on managing and surpassing patient expectations. Figure 2.16 illustrates the iron triangle.

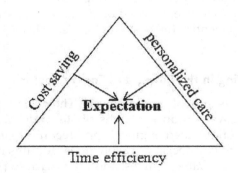

FIGURE 2.16
Iron triangle.

2.8.7.1 Impacts of Cognitive Computing in Healthcare

- Provide individuals with timely information.
- Identifies at-risk patients ahead of time.
- Predicts a patient's health demands as well as the cost of care.
- Assists in the development of customized medicine and clinical decision-making.
- Recommends a course of action based on the likelihood of success.
- Utilizes technologies such as mobile health apps and wearable sensors to improve patient involvement and communication across care settings.

2.9 Conclusions

Cognitive computing is a field that assists humans in making better judgments. To make better judgments, it uses AI techniques, big data, and advanced data analytics. We began this chapter by discussing what cognitive computing is and how it has developed from four different perspectives: discovery of knowledge, cognitive knowledge, big data, and lastly, cognitive computing. After that, there was a comparison between AI and cognitive computing. The architecture is then developed. It consists of three components: IoT, big data, and cloud computing. Then we explained cognitive computing's enabling technologies, such as reinforcement learning and deep learning. Then, the necessity of cognitive analytics was discussed. Finally, three scenarios are used to show typical applications of cognitive computing: robot technology; retail, logistics, power, and finance businesses; and medical cognitive systems. Because cognitive computing cannot replace humans, it has no effect on the job market. Instead, it aids and enables individuals to be superior in their domains.

References

1. Cognition, Wikipedia, https://en.wikipedia.org/wiki/Cognition.
2. Wang, Y, et al. Perspectives on cognitive informatics and cognitive computing. *International Journal of Cognitive Informatics and Natural Intelligence (IJCINI)* 4(1) (2010): 1–29.
3. Patel VL, Kannampallil TG. Cognitive informatics in biomedicine and healthcare. *Journal of Biomedical Informatics* 53 (2015): 3–14.
4. Daniel, J, et al. Blockchain technology, cognitive computing, and healthcare innovations. *Journal of Advances in Information Technology* 8(3) (2017).
5. Bhaskar, KB, Prasanth, A, Saranya, P. An energy-efficient blockchain approach for secure communication in IoT-enabled electric vehicles. *International Journal of Communication System* 35 (2022): 1–25.
6. Lavanya, S, Prasanth, A, Jayachitra, S, Tuned, A. Classification approach for efficient heterogeneous fault diagnosis in IoT-enabled WSN applications. *Measurement* 183 (2021): 1–22.
7. Sekar J, Aruchamy, P. An efficient clinical support system for heart disease prediction using TANFIS classifier. *Computational Intelligence* 38 (2022): 610–640.

8. Prasanth, A, Jayachitra, S. A novel multi-objective optimization strategy for enhancing quality of service in IoT-enabled WSN applications. *Peer-to-Peer Networking and Applications* 13 (2020): 1905–1920.

9. Saravanakumar, P, Sundararajan, TVP, Dhanaraj, RK, Nisar, K, Memon, FH, et al. Lamport certificateless signcryption deep neural networks for data aggregation security in WSN. *Intelligent Automation & Soft Computing* 33(3) (2022): 1835–1847.

10. Gupta, S, Kar, AK, Baabdullah, A, Al-Khowaiter, WAA. Big data with cognitive computing: A review for the future. *International Journal of Information Management* 42 (2018): 78–89, ISSN 0268-4012.

11. Lv, Z, Qiao, L. Deep belief network and linear perceptron based cognitive computing for collaborative robots. *Applied Soft Computing* 92 (2020): 106300.

12. Gudivada, VN. Cognitive computing: Concepts, architectures, systems, and applications. *Handbook of Statistics*, Vol. 35, 2016, pp. 3–38.

13. Gudivada, VN, Irfan, EF, Rao, DL. Chapter 5 – Cognitive analytics: Going beyond big data analytics and machine learning. Editor(s): VN Gudivada, VV Raghavan, V Govindaraju, CR Rao, *Handbook of Statistics*, Elsevier, Vol. 35, 2016, pp. 169–205, ISSN 0169-7161, ISBN 9780444637444.

14. Pichumani, S, Sundararajan, TVP, Dhanaraj, RK, Nam, Y, Kadry, S. Ruzicka indexed regressive homomorphic ephemeral key benaloh cryptography for secure data aggregation in WSN. *Journal of Internet Technology* 22(6), (2021): 1287–1297.

15. Wan, S, Gu, Z, Ni, Q. Cognitive computing and wireless communications on the edge for healthcare service robots. *Computer Communications* 149 (2020): 99–106.

16. Mathew, PS, Pillai, AS, Palade, V, Applications of IoT in healthcare. Editor(s): S Arun Kumar, T Arunkumar, MS Venkatesan, *Cognitive Computing for Big Data Systems Over IoT: Frameworks, Tools and Applications*, Springer International Publishing, 2018, pp. 263–288.

17. Irfan, MT, Gudivada, VN, Chapter 9 – Cognitive computing applications in education and learning, Editor(s): VN Gudivada, VV Raghavan, V Govindaraju, CR Rao, *Handbook of Statistics*, Elsevier, Vol. 35, 2016, pp. 283–300, ISSN 0169-7161, ISBN 9780444637444.

18. Appel, AP, Candello, H, Gandour, FL. Cognitive computing: Where big data is driving us. In *Handbook of Big Data Technologies*, 2017, pp. 807–850.

19. Haldorai, A, Ramu, A, Chee-Onn, C. Editorial: Big data innovation for sustainable cognitive computing. *Mobile Networks and Applications* 24(1) (2019): 221–223.

20. Aghav-Palwe, S, Gunjal, A. Chapter 1 – Introduction to cognitive computing and its various applications, Editor(s): M Mittal, RR Shah, S Roy, *Cognitive Data Science in Sustainable Computing, Cognitive Computing for Human-Robot Interaction*, Academic Press, 2021, pp. 1–18, ISBN 9780323857697.

21. Wang, Y. Cognitive computing and World Wide Wisdom (WWW+), *9th IEEE International Conference on Cognitive Informatics (ICCI'10)*, 2010, pp. 4–5, doi:10.1109/COGINF.2010.5599737.

22. Esser, SK, et al. Cognitive computing systems: Algorithms and applications for networks of neurosynaptic cores, *The 2013 International Joint Conference on Neural Networks (IJCNN)*, 2013, pp. 1–10, doi:10.1109/IJCNN.2013.6706746.

23. Ishwarappa, JA. A brief introduction on big data 5Vs characteristics and Hadoop technology, *Procedia Computer Science* 28 (2015): 319–324.

3

IoT with 5G in Healthcare Systems

S. Vijayanand

Sri Venkateswara College of Engineering

CONTENTS

3.1 Introduction

As per Brand Essence statistical surveying, the value of IoT in the medical care market is estimated to surpass $10 billion in 2024. Other significant advancements affect this development gauge. IoT is gradually acquiring a foothold and developing along with new super-quick 5G wireless mobile service, artificial intelligence (AI), and big data. Coupling of these incredible innovations with the Internet of things (IoT) will very likely change the medical care industry. IoT in medical services, for instance, utilizing 5G wireless mobile service and AI, could totally change in what way patients are observed and treated from a distant place (remotely) [1].

DOI: 10.1201/9781003256243-3

Basically, the IoT is an idea dependent on the idea of full universal figuring, which is the handling of data connected to outside movement or articles [2]. Universal processing involves permitting electronic gadgets furnished with chips and sensors to communicate with each other. Then again, actually, those electronic gadgets have Internet access; IoT is an inescapable organization. The IoT within the medical offering's enterprise is an incredible illustration of this pervasive processing. In a clinic, for instance, many electronic gadgets can be introduced to screen patients' well-being status 24 hours every day, communicate with each other, decide, and transfer information to a medical care cloud storage.

3.1.1 Examples of IoT in Healthcare

How might IoT be utilized successfully in medical services? How about we take a gander at three functional IoT medical services models underneath:

- Detecting and transferring state-of-the-art patient data to the cloud in crisis circumstances, regardless of whether from the emergency vehicle or at home.
- Clinical gadgets that can perform self-support. IoT medical services gadgets will distinguish low edges, sense their own parts, and communicate with clinical staff and makers.
- Wearables and IoT can help monitor intensely sick and old patients straightforwardly with a medical care office.
- Telemedicine is a "crude" illustration of the IoT in medical offerings.
- Through camcorders along with other electronic gadgets, a patient can be monitored and, at times, treated remotely utilizing IoT.

To see how the IoT works in healthcare, first see how the IoT functions overall. As recently expressed, an IoT part is a gadget with a sensor that can associate with the actual world and send information to the Internet. In medical services, these gadgets can gather different patient information and get input from healthcare suppliers [3]. Consistent glucose observation for insulin pens is an illustration of IoT healthcare that functions admirably for diabetic patients. These gadgets can speak with each other and, sometimes, make basic moves that could give ideal help or even save daily existence. An IoT medical services gadget, for instance, can settle on wise choices such as calling a medical care office assuming an old individual tumbles down. In the wake of gathering aloof information, an IoT medical care gadget would send these basic data to the cloud so that specialists could follow up on them, view the overall patient status, decide if an emergency vehicle is required and what sort of help is required, etc. [4]. Subsequently, the IoT healthcare can possibly essentially work on not just a patient's well-being and help with basic circumstances, yet additionally the usefulness of well-being representatives and clinic work processes.

3.1.2 Working Principles of IoT in Healthcare

How about we check out an illustration of an IoT medical care work process:

- A sensor gathers information from a patient, or a specialist or medical attendant enters information [5–7].
- An IoT gadget examination gathers information using machine learning (ML) algorithms, for example AI (ML).

- The gadget decides whether to perform or send the information to the storage in the cloud.
- In view of the information given by the IoT gadget, specialists, healthcare professionals, and even robots are able to make decisions and informed choices.

Albeit not all IoT gadgets ought to have sensors, they should have a radio set and then a TCP/IP address to communicate with others over the Internet. A gadget can be viewed as an IoT gadget assuming it has Internet access. Accordingly, every one of the smartphones is an IoT gadget. A smartphone stacked with the right medical care applications can help you in distinguishing sicknesses and working on your well-being. Applications that identify skin malignant growth utilizing your voice are one model. To plan moles on your skin, utilize a camera and AI-controlled calculations. Applications for rest, yoga, wellness, and pill the executives are some different models. Regardless, a smartphone is a smart phone. Its essential application isn't medical care checking. A devoted medical care IoT gadget can achieve undeniably more.

- **Smartwatch:** Wearables sold in customer gadgets stores incorporate a sensor just as an Internet association. Some of them (such as the iWatch Series 4) may even screen your pulse, manipulate diabetes, assist in language training, assist in act improvement, and also distinguish seizures.
- **Insulin Pumps and Continuous Glucose Monitoring (CGM):** These devices can quantify glucose degrees and send the statistics to a mobile smartphone software. Diabetes sufferers can utilize these gadgets to screen their blood glucose tiers and even send these statistics to a clinical services office.
- **Sensors for Brain Swelling:** These infinitesimal sensors are rooted on the inner side of the skull to help the experts in following life-threatening cerebrum wounds and forestalling further fatal expansion. They measure strain on the cerebrum and can break up in the body without the requirement for additional clinical mediation.
- **Sensors that Can Be Ingested:** Physician-recommended medicine is guzzled with an infinitesimal absorbable irrefutable sensor, which contains a message to a wearable recipient on the patient, which then, at that point, sends facts to a devoted cellular smartphone application. This sensor can help consultants in ensuring that the patients take their medications on a consistent basis.
- **Video Pills that Are Brilliant:** A brilliant pill can photo itself as it goes through a patient's gastrointestinal system. It would then be able to send the information gathered to a wearable gadget, which will then, at that point, send it to a devoted cell phone application (or directly to the application). Video pills can likewise support the representation of the gastrointestinal plot [6].

3.1.3 Advantages of IoT in Healthcare

The utilization of IoT in the medical care industry enjoys various benefits. The main advantage, nonetheless, is that treatment results can be fundamentally improved or augmented on the grounds that information gathered by IoT medical care gadgets is profoundly precise, permitting informed choices. Since all the patient data can be estimated rapidly and shipped off a leading body of specialists or a medical services cloud stage, well-being offices and professionals will actually want to limit blunders. Computer-based intelligence-controlled calculations running on these IoT gadgets could likewise help with settling on reasonable choices or suggestions dependent on the existing information.

One more huge benefit of IoT in medical services is cost savings. Other than that, critical patients will actually want to remain at home while different IoT gadgets screen and communicate terrifically significant data to the well-being office, bringing about less clinic stays and specialist visits. Well-being offices will actually want to further develop infection in the board by getting definite data from countless IoT gadgets. They'll have more information coming in continuously than any other time in recent memory. Regardless, this presents various difficulties.

3.1.4 Obstacles of IoT in Healthcare

In spite of the fact that IoT in medical care has numerous extraordinary advantages, there are a few difficulties that should be attended to. Without recognizing these difficulties, IoT healthcare arrangements can't be considered for execution. Monstrous measures of created information are taken care of into the framework. A great many tools in an introverted medical services workplace, in addition to 1,000 additional sending information from far away areas and all progressively will produce monstrous events of data. The information created by IoT in medical services will more likely than not increase storing necessities from terabytes to petabytes. Calculations controlled by artificial intelligence and the cloud, when used with precision, can help in detecting and organizing this information. However, this method will require a certain amount of investment to be developed. Thus, fostering a huge scope of IoT medical services arrangement will take a lot of time and exertion [8].

IoT gadgets will expand the assault surface. IoT medical services give various advantages to the medical industries; however, it additionally presents various security imperfections. Programmers could get close enough to clinical gadgets associated with the Internet and take or change the information. They can likewise go above and beyond and contaminate IoT gadgets with the scandalous Ransomware infection, tainting a whole medical clinic organization. That implies the programmers will accept prisoner patients just as their pulse screens, circulatory strain perusers, and cerebrum scanners.

The current programming foundation is obsolete. Many emergency clinics' IT frameworks are obsolete. They will forestall legitimate IoT gadget coordination. Subsequently, medical services offices are expected to enhance their IT cycles and deliver more current and up-to-date programs. They will also have to use virtualization (advanced such as SDN and NFV), just like super-fast remote and versatile organizations such as LTE Advanced and the 5G network [9].

IoT in the medical services industry can possibly further develop parts such as clinical gadgets and administrations. It can likewise further develop medical care applications such as telemedicine, patient observation, drug delivery, imaging, and in general emergency clinic work processes. It can likewise prompt the improvement of novel medicines for different illnesses. The IoT in medical care will be utilized by careful focuses, research associations, and surprisingly legislative foundations, notwithstanding clinics and offices.

The IoT in the field of medical services is certifiably not an isolated phenomenon. To help medical services offices change genuinely, all IoT gadgets and organizations would have to be associated with different innovations. As recently expressed, IoT will change the medical care industry; however, it will likewise require information, high-velocity correspondence, and just as appropriate security and consistency.

5G will provide the space and versatility the medical services industry needs for IoT. As a result, AI-controlled arrangements will include information lakes assembled from a variety of gadgets. Such AI calculations will be used in big data methodologies to continuously dissect information and install on basic welfare choices. Virtualization will contribute to the reduction or elimination of the obsolete framework within clinics. The vast majority of these innovations are as of now being utilized by IoT to assist medical care with advancing, and this pattern will just proceed. Medical services and the IoT will become indivisible as soon as possible, totally changing how we approach medical services [10].

3.2 5G Connectivity

The presentation of 5G availability gives monstrous association power, permitting medical care suppliers to give basic consideration on request. 4G organizations and other correspondence innovations are at present being utilized widely in medical services for savvy medical services applications. Notwithstanding, as the savvy medical services market develops, the interest for a steady organization association develops. Thus, brilliant medical services applications are relied upon to be upheld by the arising 5G organization. Most prerequisites, for example ultra-reliability and low dormancy interchanges or basic machine-type correspondence, high data transfer capacity, high thickness, and high energy productivity, can be met by it for countless gadgets, machines, and sensor-based applications associated with the IoT environment [11].

5G availability isn't just quicker than a 4G network; it likewise gives huge association power, permitting a great many gadgets to be associated simultaneously to foster 5G-empowered savvy medical services arrangements. Moreover, the blend of 5G organizations and IoT is relied upon to change medical care later on. It will give the framework needed to ship enormous measures of information where close ongoing information access and the capacity to settle on split-second choices are basic. 5G innovations, for example gadget-to-gadget correspondence mmWaves, the full-scale cell, and little cells, can address the ultra-densification and high-energy utilization difficulties of remote sensor networks-based IoT applications.

Little cells are low-controlled radio access hubs with a scope of a couple of meters to a mile in measurement that can be utilized for brilliant medical care arrangements that require high information rates, such as a far-off medical procedure. Little cells are ordered into three kinds, going in size from femtocells to picocells to microcells. Femtocells, then again, are utilized to expand inclusion and limit in a little region, such as an emergency clinic or a home. It has a scope of 0.1 km and can uphold up to 30 clients. 5G will have a huge effect on medical services as well as on different businesses such as assembling, auto, shrewd network, buyer gadgets, etc. In medical care, there are a few promising 5G IoT use cases. Some of them are recorded underneath.

3.2.1 5G-IoT Healthcare Applications

- **Virtual Consultations Using High-Definition (HD) Video:** Enhanced mobile broadband (EMBB) gives quicker information rates across wide inclusion regions to work on ultra-HD video counsel. It empowers essential and optional consideration experts to offer greater types of assistance, for example beginning screening evaluations, standard check-ups, treatment/restoration meetings, and, progressively, visual findings (for example recognizing dermatological conditions and indications). By leading these arrangements over the web, the patient is assured of the weight of making a trip to see medical care experts, and the expense of every arrangement is decreased [12].

- **Investigation of Real-Time Data:** The main benefit of 5G is continuous information transfer for further developed information examination. It can give super-low inertness, permitting information to be handled all the more rapidly at the organization's edge. This will help with an assortment of uses, including distant analysis, far-off medical procedures, ongoing patient well-being checking, and working on quiet commitment. It further develops information investigation abilities by giving solid availability between cell phones, versatile applications, cloud administrations, gadgets, sensors, and frameworks.

- **Far-off Patient Monitoring:** With 5G, the dependability, execution, and the ability to give proactive medical care administrations and persistent infection the board improves. The 5G organization empowers secure and quicker information to move to the cloud when sensors, wearables, and e-well-being gadgets are connected as a feature of an IoT arrangement. Man-made intelligence empowered programming in the cloud screens and examined patient vitals without expecting patients to visit essential consideration offices.

- **Conduct Recognition Using Video Analytics:** Video investigation can help with distinguishing strange conduct in patients, for example falls or occurrences in which they become a threat to themselves. 5G gives a high transfer speed, permitting HD video to be sent for handling and examination on the edge or cloud utilizing information gathered from shrewd cameras on the gadget. Dive more deeply into how 5G is set to upset an assortment of enterprises in the coming decade [13].

3.2.1.1 From Vision to Reality

In contrast to 4G, 5G correspondence innovation, related to IoT, big data analytics, AI, and ML, will fundamentally affect the medical care framework. 5G, then again, requires a critical venture. Chipsets, security, network-associated gadgets, and assets all require the mastery of 5G designing administrations and the advancement of availability stages to work out as expected. OEMs ought to work together with a believed innovation accomplice for 5G IoT application advancement to understand the maximum capacity of 5G and work on mechanical abilities to make future items.

3.2.1.2 Abilities of VOLANSYS 5G

As a one-stop arrangement supplier, VOLANSYS gives start to finish item designing administrations. We have many years of involvement offering 5G designing types of assistance and creating network stages utilizing cell advances such as GPRS and 3G/4G/LTE. This has helped our customers from an assortment of enterprises in changing their organizations into profoundly useful and trustworthy models.

3.2.2 5G Connectivity, e-Health, and m-Health

5G carries things to an unheard-of level. Networks utilizing 5G will be invested with an earth-shattering ability to do information serious exercises identified with e-well-being and mobile well-being, with a lightning velocity of more than 1 GB/second (or m-well-being). Data about your circulatory strain, pulse, blood glucose levels, and dosing examples will be gathered in the billions and transferred at always speeding up by your cell phone from sensors, screens, and other keen wearable gadgets.

Weighty advanced records containing clinical symbolism and great recordings will be fit for quick transfers that can be sent across significant distance organizations and afterward downloaded onto your facility or medical clinic's servers. All of your crucial well-being data will be measured, put away, and refreshed at each moment with unmatched proficiency on account of durable batteries and fast associations [14].

3.2.3 5G in Revolutionized Healthcare

3.2.3.1 Fast Transmission of Huge Imaging Documents

X-rays and other picture machines produce extremely huge documents that must oftentimes be shipped off to an expert for audit. At the point when the organization's data transfer capacity is restricted, transmission can consume most of the day or flop completely. This implies the patient should wait much longer for treatment, and suppliers will actually want to see less patients in a similar measure of time. Adding a rapid 5G organization to existing designs can help rapidly and dependably transport huge information records of clinical symbolism, working on both admittances to mind and mind quality.

The PET scanner at the Austin Cancer Center produces amazingly enormous documents – up to 1 gigabyte of information for every quiet per study. "You must have the organization execution to deal with that much information from one part of town to different," says Jason Lindgren, CIO of Austin Cancer Center. "Already, we needed to send the documents late at night. The review is currently in progress when the patient leaves the scanner. Specialists benefit from this on the grounds that they can get the outcomes they need quicker" [15,16].

3.2.3.2 Broadening Telemedicine Administrations

As indicated by Market Research Future, the telemedicine market will develop at a yearly development pace of 16.5% somewhere in the range of 2017 and 2023. As indicated by the review, the justification for the anticipated increment is the expanded interest for medical care in provincial regions, just as an increment in government drives.

3.2.3.3 Enhancing Augmented Reality, Virtual Reality, and Spatial Computing

While increased reality (AR), augmented reality (VR), and spatial figuring are as of now being utilized in medical services, 5G may ultimately work on a specialist's capacity to convey imaginative, less intrusive therapies. Among the numerous extreme expected applications for 5G, one of the most invigorating is its part in recreating complex clinical situations and empowering elective medicines for the basically sick. AT&T is at the cutting edge of this intriguing field, researching ways of applying 5G to clinical difficulties. AT&T

is working with VITAS® Healthcare to examine the effect of future 5G-empowered AR and VR on quiet commitment. The objective is to give quieting, diverting substance by means of 5G-empowered AR and VR to at death's door patients in hospice to diminish torment and tension.

3.2.3.4 Predictable, Constant Remote Observing

Medical care suppliers can utilize IoT gadgets to screen patients and gather information that can be utilized to work on customized and preventive consideration. Wearables, a typical kind of remote checking, are said to build patient commitment with their own well-being by 86% of specialists, as indicated by Anthem. Moreover, wearables are relied upon to diminish emergency clinic costs by 16% over the course of the following 5 years. Notwithstanding the advantages, the utilization of remote checking innovation is restricted by the organization's ability to deal with the information.

Slow organization speeds and temperamental associations might keep specialists from acquiring the continuous information they need to settle on fast medical services choices. Medical care frameworks can give remote observing to more patients utilizing 5G innovation, which has lower dormancy and higher limits. Suppliers would then be able to be certain that they will get the information they need continuously and that they will actually want to give the consideration their patients require.

3.2.3.5 ML & AI

Numerous basic medical care capacities are starting to utilize man-made brainpower (AI) to decide expected conclusions and decide the best therapy plan for a particular patient. Besides, AI can assist in foreseeing which patients are bound to encounter post-employable entanglements, permitting medical care frameworks to intercede prior when important. Due to a lot of information needed for ongoing fast learning, super-solid and high-transfer speed networks are required. Besides, suppliers oftentimes require information access from their cell phones [17].

3.3 Internet of Medical Things (IoMT)

The IoMT is a reasonable use of gadgets from the IoT combined with MedTech devices used in the clinical field. These gadgets, otherwise called web-associated clinical gadgets, gather, make due, cycle, and store clinical information. Medical services experts can screen patients' critical biometrics progressively by means of emergency clinic networks on account of associated gadgets' capacity to gather and send information. These correspondence advances give various benefits to medical services suppliers, some of which are not promptly clear [18,19]. This innovation, by remotely sending information, empowers specialists to get to clinical information in distant areas and monitor any potential issues that might emerge, subsequently assisting in forestalling future entanglements while performing more exact judgments.

Expect a patient is told not to gauge in excess of a specific sum or, more than likely, medical conditions will emerge. A patient can without much of a stretch monitor this critical measurement with the assistance of an application utilizing a clinical gadget, for example an IoMT scale or other web-associated gadgets. Given the patient's information and condition, the application can likewise share data on solid weight control plans or be utilized as a glucose screen and movement tracker. Utilizing pop-up messages can assist in setting off inspiration and conducting changes. It can likewise impart relevant data to clinical faculty, so they know about how the patient has advanced before emergency clinic visits. So, IoMT programming applications can work on the productivity of cycles. They can be utilized for distant patient checking on the grounds that they empower constant information trade between point-of-care gadgets and clinic organizations. This results in more precise conclusions while diminishing the quantity of in-person clinical visits.

IoMT is an assortment of gadgets and clinical applications targeted at healthcare data innovation frameworks through web-based PC organizations. Clinical gadgets with Wi-Fi allow the machine-to-machine correspondence, which is the implementation of the IoMT. IoMT gadgets speak with cloud steps such as Amazon Web Services, where captured information is stored and examined. IoMT is additionally alluded to as medical care IoT. Far-off tolerant observing of individuals with ongoing or long-haul conditions, following patient prescription orders and the area of patients, conceded to emergency clinics, and patients' wearable mHealth gadgets that can send data to parental figures are altogether instances of IoMT.

Clinical gadgets that can be changed over to or conveyed as IoMT innovation incorporate mixture siphons that are associated with investigation dashboards and clinic beds equipped with sensors that action patients' important bodily functions. Likewise, with bigger IoT, there are currently more expected uses of IoMT than any other time in recent memory in light of the fact that numerous shopper cell phones incorporate near-field communication (NFC) radio frequency ID (RFID) labels that permit the gadgets to impart data to IT frameworks. RFID labels can likewise be joined to clinical gear and supplies, permitting emergency clinic staff to monitor how much stock they have available.

3.3.1 IoT Security

Telemedicine is the act of remotely checking patients in their homes utilizing IoMT gadgets. This sort of therapy wipes out the requirement for patients to visit an emergency clinic or doctor's office at whatever point they have a clinical inquiry or an adjustment of their condition. The security of touchy information that goes through the IoMT, for example, ensured well-being data administered by the Health Insurance Portability and Accountability Act, is a developing concern [20]. The IoMT can likewise support the initiation of individual crisis reaction frameworks and the administration of constant infections. Consider savvy gadgets that can assist you in checking your glucose levels and pulse. Regardless of whether you are in a distant area, you can utilize wearable gadgets to impart movement tracker information to a far-off clinical benefit supplier and get a clinical-grade analysis.

For this model, think about patient X. Patient X, a moderately aged lady with a diabetes family ancestry, awakens one day feeling sick. Her primary care physician had luckily given her a shrewd glucometer. Dissimilar to a standard glucometer, this exceptional clinical gadget can interface with the Internet. It permits patient X to effortlessly follow her glucose levels, yet it doesn't stop there. Since the gadget is an IoMT-empowered glucometer, the specialist can remotely check her glucose levels exhaustively while keeping away from any in-person clinical visits. Care suppliers might have the option to decide when patient X is encountering a risky glucose level in the wake of auditing the information, and in light of the fact that they know about her clinical history, a customized arrangement can be suggested [21].

3.3.2 IoMT Obstacles

Altogether, for IoMT advances to work appropriately, important information should be gathered. Indeed, with all IoT advancements, IoMT doesn't work as expected without legitimate information. This isn't shocking given that one of the essential advantages of this innovation is that it helps with navigation. IoMT clinical gadgets should permit information to be gathered, handled, and utilized. This requires something beyond introducing a signal to gather information. It should likewise be changed over into valuable data that clinical faculty can use to simply decide. This is presently conceivable on account of MedTech applications and IT frameworks [22–24].

Applications are assuming a huge part in the arrangement of IoMT advances. It is futile to have incredible gadgets assuming the product that controls them is deficient. MedTech applications intended to work with IoT gadgets should furnish clients with satisfactory degrees of constancy. In this sense, quality confirmation is a necessity that these sorts of applications can't overlook. With regard to guaranteeing patient security and well-being, quality is an unquestionable requirement. Security is a basic achievement factor for any IoMT innovation. This alludes to something beyond quiet security, which is basic. It ought to be the first concern to guarantee it. Regardless, online protection, especially as it identifies with information security, ought to be a main concern. Clinical data are incredibly touchy, so clients should be guaranteed that their information is secured as per HIPAA guidelines [25].

3.3.3 Opportunities in Healthcare Information Technology

In light of our perceptions, there are two significant manners by which web-associated clinical gadgets can help patients. Customized patient care medicine much of the time seems to need compassion for patients. This is because of the way that well-being frameworks should normalize strategies to lessen costs and keep up with effectiveness. This affects everything from medicines to results. This doesn't need to be the situation [26]. IoMT information empowers an all the more amicable, gentler type of customized medication that considers patients' real necessities. Rather than taking a one-size-fits-all methodology, medical services experts would now be able to think about individual necessities with IoMT [27].

3.3.4 Computerized Therapeutics (DTx)

Perhaps the most engaging chance for medical care organizations hoping to execute IoMT gadgets is computerized therapeutics (DTx). It is a part of medication that has filled in ubiquity lately, depending on proof-based helpful intercessions to change patient conduct. To achieve this, DTx utilizes programming to forestall, make due, or treat a particular disease. DTx procedures can foster better far-off associations with patients on account of the force of IoMT gadgets. We guess that as this kind of treatment and IoMT gadgets gain fame, so will their separate applications [28].

Applications of MedTech are at the core of the worldwide IoMT market. I've as of now referenced how even the most remarkable clinical gadgets are delivered futile without an amazing application to back them up. The most ideal way to guarantee that this doesn't turn into an issue is to look for the administrations of an accomplished clinical expert. With innovation, individuals' lives are in question, and the eventual fate of medical services frameworks needs engineers and architects who can comprehend its complexities to help clinical information work for you. IoMT application improvement isn't something to leave in the possession of anybody. All things considered, what is IoMT useful for on the off chance that you don't have the right programming supporting your organization of associated gadgets [29,30].

3.4 Conclusions

The IoMT is made possible by 5G. 5G has a few key properties that will empower a huge organization of associated "things" – gadgets and machines that can speak with each other, with or without human mediation. 5G empowers the "Internet of Things" by giving significantly further developed versatile broadband information rates, taking into consideration at any point quicker streams of bigger measures of data.

- Incredibly low idleness and dependability – appropriate for crucial administrations.
- The capacity to scale fundamentally and productively to interface countless sensors.
- Further developed security, for example biometric recognizable proof capacities that assist in ensuring data trustworthiness.

Clinical gadgets, wearables, far-off sensors, and remote fixes that screen and electronically send indispensable signs, actual work, individual security, and medicine adherence are instances of IoMT gadgets. 5G is an especially viable IoMT impetus. Due to 5G's pervasiveness, ultra-dependability, and capacity to help higher-speed transmission at much lower inactivity than the present versatile organizations, it won't just empower quicker and bigger information streams; however, it will likewise consolidate back-end server farms, cloud administrations, and remote document servers into a computational behemoth. There will be edge processing, which implies that calculations can be performed near the source, on the gadget or sensor itself, or in the cloud, contingent upon the prompt need. These 5G advancements will empower applications to handle content rapidly and give a close ongoing and responsive experience.

To put it plainly, 5G advancements do definitely more than essentially transfer pieces of information at consistently speeding up. All things considered, the "computational behemoth" portrayed above empowers the organization to brilliantly absorb and deal with a lot of information, permitting it to be changed over once again into customized proposals and activities for patients and guardians. Moreover, these 5G developments will make it simpler to share those data (by means of distributed computing).

The far and wide accessibility of 5G empowers the expansion of associated "clinical things." This "omnipresence" property emerges on the grounds that 5G is something other than an augmentation of existing 3G and 4G organizations. It joins Wi-Fi and cell networks into a solitary, consistent organization. This universality or consistency is the thing that drives the multiplication of associated gadgets and empowers, for instance, non-stop tolerance checking. The security properties of 5G are basic to guaranteeing the security and honesty of data, bringing a basic boundary down to data dispersal and use.

References

1. Islam, S. M. et al. The internet of things for health care: A comprehensive survey. *IEEE Access* 3, 678–708. doi: 10.1109/ACCESS.2015.2437951, 2015.
2. Farahani, et al. Towards fog-driven IoT eHealth: Promises and challenges of IoT in medicine and healthcare. *Future Generation Computer Systems* 78. doi: 10.1016/j.future.2017.04.036, 2017.
3. Sekar, J., Aruchamy, P. An efficient clinical support system for heart disease prediction using TANFIS classifier, *Computational Intelligence* 38, 610–640, 2022.
4. Jing, et al. Security of the internet of things: Perspectives and challenges. *Wireless Networks* 20, 2481–2501. doi: 10.1007/s11276-014-0761-7, 2014.
5. Bhaskar, K. B., et al. An energy-efficient blockchain approach for secure communication in IoT-enabled electric vehicles. *International Journal of Communication System* 35, e51891, 2022.
6. Ramakrishnan, V., Chenniappan, P., Dhanaraj, R. K., Hsu. Bootstrap aggregative mean shift clustering for big data anti-pattern detection analytics in 5G/6G communication networks. *Computers & Electrical Engineering* 95, 1–17, 2021.
7. Prasanth, A., Jayachitra, S., A novel multi-objective optimization strategy for enhancing quality of service in IoT-enabled WSN applications. *Peer-to-Peer Networking and Applications* 13, 1905–1920, 2020.
8. Gao, et al. Application and effect evaluation of infusion management system based on internet of things technology in nursing work. *Studies in Health Technology and Informatics* 250, 111–114, 2018.
9. Dac-Nhuong, L., Souvik, P. *IoT: Security and Privacy Paradigm*. CRC Press. ISBN: 9780367253844, 2020.
10. Anmulwar, S., Gupta, A.K., Derawi, M. Challenges of IoT in healthcare IoT and ICT for healthcare applications. *EAI/Springer Innovations in Communication and Computing*. Springer, Cham. doi: 10.1007/978-3-030-42934-82, 2020.
11. Gamage, R., Rashmika, M. A review on applications of internet of things (IOT) in healthcare. *Journal of the American Society for Information Science and Technology* 2020.
12. Andrews, J. G. et al. What will 5G be? *IEEE Journal on Selected Areas in Communications* 32(6), 1065–1082. doi: 0.1109/JSAC.2014.2328098, 2014.
13. Anand, A., et al. An efficient CNN-based deep learning model to detect malware attacks (CNN-DMA) in 5G-IoT healthcare applications. *Sensors* 21(19), 6346. doi: 10.3390/s21196346, 2021.
14. Latif, S., et al. How 5G wireless (and concomitant technologies) will revolutionize healthcare? *Future Internet* 9, 93. doi: 10.3390/fi9040093, 2017.
15. Akpakwu, G.A., et al. Survey on 5G networks for the Internet of Things: Communication technologies and challenges. *IEEE Access* 6, 3619–3647. doi: 10.1109/ACCESS.2017.2779844, 2017.
16. Wang, D., et al., From IoT to 5G I-IoT: The next generation IoT-based intelligent algorithms and 5G technologies. *IEEE Communications Magazine* 56, 114–120. doi: 10.1109/MCOM.2018.1701310, 2018.
17. Dananjayan, S., Raj, G.M. 5G in healthcare: How fast will be the transformation? *Irish Journal of Medical Science* 190, 497–501. doi: 10.1007/s11845-020-02329-w, 2021.
18. Saravanakumar, P, Sundararajan, T. V. P., Dhanaraj, R. K., Lamport certificateless signcryption deep neural networks for data aggregation security in WSN. *Intelligent Automation & Soft Computing* 33, 1835–1847, 2022.

19. Joyia, et al. Internet of medical things (IoMT): Applications, benefits and future challenges in healthcare domain. *Journal of Communications* 12, 2017, 240–247. doi: 10.12720/jcm.12.4.240-247.

20. Suresh, et al., *Internet of Medical Things (IoMT) – An Overview*. 101–104. doi: 10.1109/ICDCS48716.2020.243558, 2020.

21. Algarni, A. A survey and classification of security and privacy research in smart healthcare systems. *IEEE Access* 7, 101879–101894, doi: 10.1109/ACCESS.2019.2930962, 2019.

22. Patnaikm, et al. A systematic survey on IoT security issues, vulnerability and open challenges. In *Intelligent System Design*, pp. 723–730. Springer, Singapore, 2021.

23. Dwivedi, R., et al. Potential of internet of medical things (IoMT) applications in building a smart healthcare system: A systematic review. *Journal of Oral Biology and Craniofacial Research*, doi: 10.1016/j.jobcr.2021.11.010, 2021.

24. Karmakar, K.K., et al. Towards a security enhanced virtualised network infrastructure for internet of medical things (IoMT). *Proceedings of the 2020 6th IEEE Conference on Network Softwarization (NetSoft); Ghent, Belgium* 29, 257–261, 2020.

25. Papaioannou, M., et al. A survey on security threats and countermeasures in internet of medical things (IoMT). *Transactions on Emerging Telecommunications Technologies* 23, e4049. doi: 10.1002/ett.4049, 2020.

26. Pradhan, B., Bhattacharyya, S., Pal, K. IoMT-based applications in healthcare devices. *Journal of Healthcare Engineering* 6632599. doi: 10.1155/2021/6632599, 2021.

27. Dang, L.M., et al. A survey on internet of things and cloud computing for healthcare. *Electronics* 9(7), 768. doi: 10.3390/electronics8070768, 2019.

28. Sethi, P., Sarangi, S. Internet of things: Architectures, protocols, and applications. *Journal of Electrical and Computer Engineering* 1–25, 9324035. doi: 10.1155/2017/9324035, 2017.

29. Dimitrov, Dimiter, V. Medical internet of things and big data in healthcare. *Healthcare Informatics Research* 22(3), 156–163. doi:10.4258/hir.2016.22.3.156, 2016.

30. Dang, A., Arora, D., Rane, P. Role of digital therapeutics and the changing future of healthcare. *Journal of Family Medicine and Primary Care* 9(5), 2207–2213. doi: 10.4103/jfmpc.jfmpc_105_20, 2020.

4

Communication Protocols for IoMT-Based Healthcare Systems

N. Pushpalatha and P. Anbarasu
Sri Eshwar College of Engineering

A. Venkatesh
Dr. Mahalingam College of Engineering and Technology

CONTENTS

DOI: 10.1201/9781003256243-4

4.1 Introduction

IoMT is a mashup of medical devices and networking technologies. Protocols for communication are used to establish interactions between communicating entities in order to share or transfer information via any variation of a measurand. The protocol specifies the communication rules, syntax, semantics, synchronization, and various error recovery mechanisms [1]. Hardware, software, or a mix of the two may be used to implement protocols. Technology innovation and the development of IoMT pave the way for the eradication and minimization of pandemic outbreaks because most infectious diseases caused by bacteria, viruses, fungi, or parasites can be identified and their spread limited; if necessary, precautions are taken to force health systems to implement new therapy regimens that require remote monitoring.

IoMT has also established a presence in several segments such as on-body, in-home, community, in-clinic, and in-hospital. These segments include a variety of smart devices and systems that are linked together to facilitate communication and improve patient service and comfort. Even nowadays, remote telemedicine, health information technology, electronic health record, virtual doctor, 3D organ printing, self-treatment for allergies, virtual ICU, inpatient telehealth monitoring through specially equipped smart gadgets, robotic surgery, immunotherapy, etc., are possible because of technological advancements. For performing all such tasks, conversion of physical quantities and communication between the smart devices are mandatory, and the system should have a sense of communication by utilizing the communication protocols to make things simple and better. Additionally, technology can help decrease needless hospitalizations and the load on healthcare systems by linking patients to their physicians and enabling the safe transfer of medical data.

4.2 Architecture of IoT

IoT will connect a large number of items; it will necessitate a layered architecture that is adaptable. Data bandwidth should be sufficient to prevent congestion as the number of objects increases. The most typical paradigm is a three-layered architecture that comprises the perception, network, and application levels. Traditional architectures have a number of issues that new formats have worked hard to eliminate. Three-tier, middleware, service-oriented architecture (SOA), five-tier, and cloud-specific are the various types of IoT architectures [2].

4.2.1 Three-Layer Architecture

It is structured with three layers: perception, network, and application. (a) Perception is a physical layer, and it is composed of sensors for perceiving and collecting data about the environment. In this case, it is able to recognize certain types of physical things or recognize other intelligent objects in its environment. (b) The network is the intermediate layer, also called the transmission layer, that is responsible for connecting or transmitting and processing information from the perception layer's sensors linked to various communication devices, intelligent objects, and servers. Additionally, its capabilities are employed to send and process sensor data [3]. (c) The application layer takes care of giving users application-specific functionality. It defines a number of things, including remote health access and smart grid data transfer [4].

4.2.2 Middleware Architecture

Middleware is fundamentally software or a group of sublayers that exist between the technology and application levels [5]. It's necessary for interconnecting assorted areas of the network and serving as a conduit for information transfer between different LANs. It addresses reliability, scalability, and coordination, among other issues.

4.2.3 Service-Oriented Architecture (SOA)

Its goal is to improve the coordination of services' workflow while guaranteeing that software and hardware components are reused. The four tiers of SOA that communicate with one another are the perception, service, network, and application layers. The service layer is segmented into two tiers: service composition and administration. The remaining layers, with the exception of the service layer, work in the same way as in the preceding architecture models. The application layer depends on the service layer for functionality. The service layer consists of composition, administration, discovery, and service interface, with service discovery being used to locate waning requests. Service management is in charge of each item, service composition is in charge of talking with linked devices and adding or deleting items as needed, and service interface is in charge of providing a communication channel between all services.

4.2.4 Five-Layer Architecture

(a) Perception layer: This layer is the OSI model's base layer, similar to the physical layer. It includes data collection techniques such as RFID and sensors, as well as two-dimensional objects that collect data on factors such as weight, temperature, and humidity. (b) Network layer/object abstraction layer: This layer communicates with higher layers via secure links. GSM, 3G, Wi-Fi, W-wire, and other networking protocols are used to send data to the middleware information processing unit. (c) Middleware/service management layer: This layer ensures optimal software interaction by acting as a bridge between application and perception layer sensors. (d) Application layer: This layer provides the consumer with all data comprehension as well as the various application services that they require. (e) The business/management layer ensures that the remaining four layers are checked and managed. It manages all IoT services and apps and provides high-level analytical results. This layer makes it easier for higher authorities to make the best judgments possible.

4.2.5 Cloud-Specific Architecture

It blends visualization with communication, memory, and processing to give the extensibility to meet the diverse and sometimes conflicting needs of various fields.

4.3 Architecture of IoMT

This section summarizes existing IoMT architectures in near-chronological order. Finally, the section discusses the IoMT architecture's key characteristics. It has divided the IoT design into two components based on SOA. User programs run on the upper end of this architecture/framework, and incompatible devices are made available via an object abstraction layer [6]. They used the Serial Global Trade Item Number (SGTIN-96) encoding technology for electronic product codes (EPCs) to identify the physical products. EPCs are converted to URLs and service names using the Object Name Service (ONS) from the obtained Naming Authority Pointer (NAPTR) information. This increases the accessibility of e-health services. Santos et al. discussed the use of RFID technology to enable interior location awareness and guidance in a variety of situations where Global Positioning System (GPS) technology is ineffective for facilitating the deployment of IoT within hospitals [7].

The architecture of the five-layer enables the system to perform sensing, transmission, service, application, as well as business and general management of IoT systems [8–10]. IoT implementation in IoMT includes RFID with multi-agent solution systems, which permits a flexible coupling of different components and their information among remote users. It recreated the usage of several network channels, including SMS, MMS, Bluetooth, HTML, and Wi-Fi, as well as a variety of data formats, including value, parameter, CSV, XML, and SOAP. Xu et al. emphasized the importance of a ubiquitous data access technique (UDA-IoT) in the clinical sector to manage the varying nature of IoT data [11]. They deployed UDA-IoT via Representational State Transfer (RESTful) APIs in an emergency medical decision support system. The RESTful concept is an information-oriented architecture that allows web interoperability and information management by using Uniform Resource Identifiers (URIs) to access web services.

A system architecture on the basis of the CoAP (Constrained Application Protocol) has been developed [12]. It is based on the RESTful system and adopted the IEEE 11073 guidelines for communication between personal health devices (PHDs). They explained how to use this new architecture to link the home network and the Internet, enabling the connected health vision to be realized through the use of Universal Plug and Play (UPnP) technology [13]. They created an ontology data model to authenticate the treatment procedure [14]. They also created an ontology-based patient monitoring system. A data model based on ontologies facilitates information exchange by integrating context-sensitive data and using automated reasoning to assess the contextual scenario. In studies [15,16], the authors described three distinct architectures for IoT-enabled healthcare systems: sensor networks, gateways, and cloud data centers. The gateway, which serves as a link between the sensors and the cloud layer, can communicate with and process data from a variety of sensor devices. They developed UT-Gate, an intelligent e-health gateway. A collection of these conduits was employed to build an intermediary tier analogous to fog computing in order to establish an early warning mechanism to monitor patients with a serious sickness.

An IoMT platform would be an intelligent system that consists primarily of smart devices for acquiring sensor information from a patient, an information gathering unit for sensor data processing, a network device for transmitting the clinical information across the system, a permanently or temporarily conserved unit, and a system for visualization with artificial intelligence (AI)-based schemes for making physician-friendly decisions [17,18], and the IoMT architecture model is illustrated in Figure 4.1.

4.4 IoMT Communication Protocols

Many IoMT standards have been proposed to make applications easier to use [19]. The guidelines are striving to become the industry standard for connected devices. The best

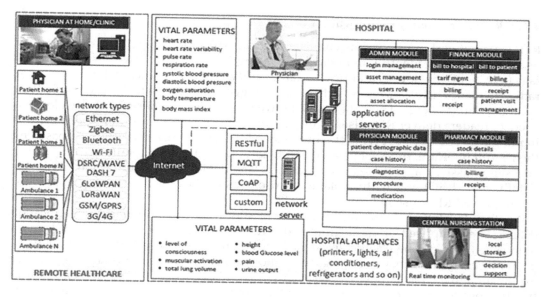

FIGURE 4.1
Architecture of IoMT.

Layers	Protocols
MAC Layer	Zigbee, RFID, TSMP, NFC, Bluetooth IEEE 802.15.1, Wireless HART, Weightless, Wi-Fi
Application Layer	COAP, MQTT, AMQP, DDS, XMPP, Web sockets

FIGURE 4.2
Protocols of IoMT.

protocol for each application will be different. When comparing existing wireless technologies, the choice is not obvious because each offers its own set of benefits and drawbacks. As a result, new protocols with features tailored to the demands of connected objects have arisen, such as low battery consumption, long range, low throughput, and ease of implementation. The classification of practically all protocols based on the main layer of the IoMT architecture is provided in the parts that follow (Figure 4.2).

4.4.1 Zigbee Protocol

The Zigbee Alliance is in charge of keeping the standard up to date. Its purpose is to make WPAN construction easier and less expensive when compared to traditional technologies such as Bluetooth [20] or Wi-Fi. It is a limited-range radio wave protocol called wireless personal area network (WPAN). Zigbee/RF4CE provides significant benefits in complex systems by merging economic-power operation with increased safety, resilience, and scaling factors when a greater number of clusters are involved. This protocol is also appropriate for IoMT and M2M implementations that require wireless sensor networks (WSNs) [21].

4.4.2 Radio-Frequency Identification (RFID)

RFID is a network-connected mechanism that consists of a group of technologies that use radio waves to identify and detect goods (tags). It consists of two sorts of devices: tags (a recognizing device) and readers (an identifying device). The reader device emits radio-frequency (RF) waves, which activate tagged devices, which react with their identification (ID) tags. The data transit between them is controlled by the readers. As needed, readers deliver RF pulses to interrogate the tags in the area. In response to this inquiry, tags react by supplying their tag IDs. RFID systems are characterized based on their radio interface, frequency band, bandwidth, tag autonomy (totally active, passive, and semi-passive), and standard compliance. The deployment of ultrahigh frequency (UHF) smart RFID tags with implanted sensors, as well as the downsizing of readers, has improved RFID use in IoT networks [22].

Wristbands, clothes, footwear, and other objects with radio-frequency identification (RFIDs) are comprised of an antenna and a tiny microchip that is incorporated into a mono-shell that can be electrically identified. When readers broadcast an information-gathering radio-frequency pulse, tags use radio frequencies to communicate their identity information to the reader devices. Despite the lack of line of sight (LOS), this communication is dependent on the tag's proximity to the reader device. The range of transmission is governed by the type of used equipment.

4.4.3 Time Synchronized Mesh Protocol (TSMP)

TSMP is a wireless protocol incorporated with auto-organizing devices. It utilizes the sensor nodes in WSN called motes. SMP synchronization is possible, and communication between motes occurs in time slots such as TDMA. TSMP is intended to function in noisy areas and avoid interference through channel hopping. It is generally used in applications that need dependability and durability.

4.4.4 Near-Field Communication (NFC)

It aids devices to communicate in extremely close proximity (in the range of a few centimeters at most) because it has been widely adopted by mobile device suppliers, making it available to the general public for a variety of purposes [23]. ECMA-340 and ISO/IEC 18092 already have standards for NFC. Unlike Bluetooth, it uses low-power lines for transmission and does not require pairing. Simply bringing one gadget near to another enables communication. This function needs the client to interact with the gadget while it is in use. It is a method of ensuring the safety and security of technology usage, as the device's facility is only functional when the owner is there.

Due to the fact that NFC devices have the ability to act as tags and readers, their functionality is analogous to that of RFID. At the 13.56 MHz frequency, communication occurs in either an active or passive mode. It typically operates within a range of around 0.2 m of another device, is sensitive to near-fields and even contact, and transmits at a rate of 424 Kbps. The active device initiates the link in passive mode communication by generating a wave carrier that initiates the functioning of the passive devices by creating the link. Thus, the passive device modifies and transmits its data via this carrier. Both the initiating and receiving devices interact in an active mode by producing their own carrier waves. These devices require peripheral power sources to operate.

NFC tags and readers support three distinct modes of operation: card emulation (CE) mode, reader/writer (R/W) mode, and peer-to-peer or point-to-point (P-P) mode. Typically, in CE mode in NFC, the tag is read by the active device, which is a kind of passive device. Both of these gadgets can be active or passive. Two nodes are linked in an ad hoc or peer-to-peer fashion to do data exchange in the NFC, P-P mode stated by ISO/IEC 18092. Due to the widespread use of NFC as a supplementary technology, it has developed into a significant technical solution for a variety of smart products, smart wearables, pay machines, and other gadgets.

In applications such as the following that require people to provide individualized services and information, particularly those found in e-health, these technologies are frequently employed [25]. This perspective presents the unusual concept of "object" or "thing" socialization, which has never been seen before. Under this paradigm, the connection between the transferred object and the person holding it creates a one-of-a-kind co-ownership and relationship between the two parties. This connection could have an impact on human-environment interactions by influencing how people make decisions and raising the level of consciousness of objects.

The RF transmission data rate over the physical layer of NFC is 106, 212, 424, or 848 Kbps, based on the modulation and coding scheme type used. Message exchange, connection establishment, anti-collision bit transfer, emulation modes, activation processes, and data delivery are all handled by the NFC MAC layer. Figures 4.3 and 4.4 illustrate these functions [24].

4.4.5 Bluetooth IEEE 802.15.1

It is a widely used wireless system based on the IEEE 802.15.1 specification. Its basic rate (BR) is a 2.4 GHz short-range wireless networking protocol that is widely used around the world. Conventional Bluetooth's benefits include low battery usage, ease of setup, and compatibility for a star network topology with an infinite number of nodes [26]. It comprises the v1.0 and v1.0B versions, which have voice dialing, call mute, redial of the previous number, a range of 10 m, and version v1.2 with dynamic frequency hopping. The (Enhanced Data Rates) EDR v2.0+ and +EDR v2.1 versions included features such as greater resistance to the impact made by radio waves, increased indoor coverage, and a 100 m line of sight (LOS) range. Approaches with high transmission rates and less power usage were also enhanced in the +EDR versions v2.0 and v2.1. The EDR v2.1 employs a sniffing mechanism to mitigate interference by allowing fewer broadcasts to access the medium.

4.4.6 Wireless-HART

According to the IEEE 802.15.4 specification, it is a centralized wireless network. It was developed to satisfy the intents of wireless industrial implementations that require precise timing parameters, high levels of security, and a high level of obstacle interference. While this protocol adheres to the same standards as the IEEE 802.15.4 PHY, it utilizes TDMA technologies to create its own MAC layer. It works in the unlicensed ISM band of 2.4–2.4835 GHz, with 16 channels of each having a bandwidth of 2 MHz, and uses the physical layer of IEEE 802.15.4 as its physical layer protocol. These channels operate with a 5 MHz gap among neighboring IEEE 802.11b/g channels. The channels may support data transfer speeds of up to 250 Kbps, and the channels are numbered 11–26. Wireless-MAC HART's layer implements its own TDMA protocol with synchronized time slots of 10 ms.

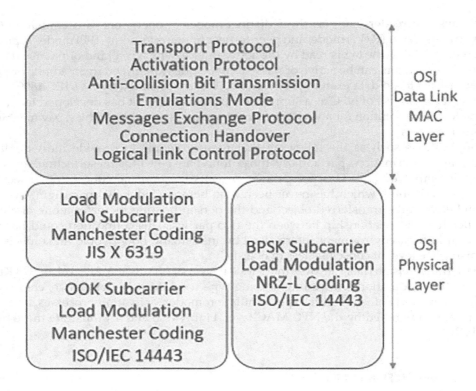

FIGURE 4.3
Communication protocol stack of NFC passive to active device.

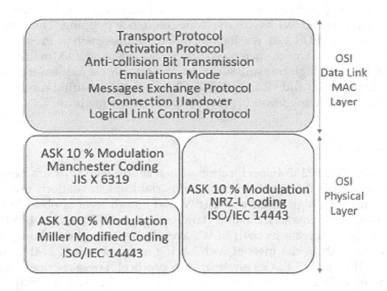

FIGURE 4.4
NFC passive to active device communication protocol stack.

These properties enable messages to transit over network architecture and interference hurdles. The network layer's support for self-healing and self-organizing mesh networking strategies enables this. While Wireless-HART is primarily a centralized wireless network, it incorporates a network administrator into its stack to control communication and routing schedules. This achieves a balance between system performance and the requirements of industrial wireless applications on the network. It is primarily concerned with mono-hop communication, whereas the network layer is responsible for network device proximity allocation [27–29].

In contrast to IEEE 802.15.4, Wireless-MAC HART's layer makes use of a time-synchronized TDMA approach mixed with frequency hopping, rather than a single frequency, which enables many devices to transmit data simultaneously over several channels. During the device's joining procedure, the network management distributes to devices communication links and channel hop sequences. Additionally, it is in charge of approving or shutting down channels that are regularly accessed and subjected to excessive levels of interference, a function referred to as the "channel blacklist." Petersen Wireless-HART, wHart-n-802-15-4e, 2011.

The eight device categories described by gateways, routers, network managers, adapters, network security mechanisms, network field devices, and network access points are all part of the Wireless-HART system, which operates in a mesh architecture [30]. They enable the deployment of features that aid in the development, maintenance, and routing of data and signaling while maintaining a minimal level of dependability. Figure 4.5 depicts a correlation of the Wireless-HART layer features and the OSI reference structure, as well as the major properties of both models.

Another part of Wireless-HART that can be accessed is the information blocks that are kept in the storage memory of each device on the network. A neighbor data block comprises information about the nodes in the immediate vicinity and the next accessible device. The block information is used to establish a connection to the network layer, and data are added to the routing table of the network layer. To achieve network synchronization while employing TDMA as a medium access mechanism, network devices must conform to extremely rigorous timing restrictions. This is because synchronization happens during both joining and regular operations.

4.4.7 Weightless

Weightless is a collection of LP-WAN protocols optimized for low-bit-rate wireless communication systems. Weightless is available in three configurations: Weightless-P, Weightless-N, and Weightless-W. The Weightless Special Interest (Weightless SIG) Group has regulated these technologies [31]. It is a network comprised of end devices (EDs) and base stations (BSs). It is a typical star topology system. Sensor nodes, sometimes referred to as leaf nodes, are referred to as EDs, whereas BSs focus on communication on EDs. The base station networks (BSNs) are comprised of the interconnections between the base stations and are responsible for system functions such as authentication, roaming, radio resource allocation, and scheduling.

Weightless is composed of two physical layers: one dedicated to high-speed data transport and another to low-speed data transfer. The physical layer for downlink in both cases is composed of functional blocks such as forward error correction (FEC) encoding, whitening, interleaving, spreading factor used, quadrature amplitude modulation (QAM) or phase-shift keying (PSK) modulation type of control, root-raised-cosine (RRC) pulse shaping, cyclic prefix insertion, and sync insertion. FEC rates, modulation types, and spreading

OSI Model	Wireless HART Layers Features
Application	Command Oriented
	Predefined data
	Types and Application Procedures
	Data Fragmentation Reassemble
Transport	Auto segmented Transfer of Large Data sets
	Reliable Stream Transport
	Negotiated Segment Size
	Transactions with or not ACKS
Network	Power Optimized
	Redundant Path
	Self – Healing Mesh Network
	Graph and Source Routing
MAC	Frequency Hopping
	TDMA slots 10ms
	Black list channels
	Security
Physical	IEEE 802.15.4. Radio
	2.4 GHz License free
	10dBm Transmission Power
	Operation Frequencies

FIGURE 4.5
Stack of the Wireless-HART protocol.

factor parameters all influence the ultimate transmission rate. The data transfer rate can be varied between 125 Kbps and 16 Mbps. A (p/2) BPSK modulation with an FEC rate and spectral scattering is being used to achieve a data rate of 125 Kbps. Additionally, when 16-QAM is used without the FEC method, 16 Mbps is obtained. When the scattering factor spectral is reduced, 16 Mbps is achieved.

The interleaving module may or may not be present, depending on the FEC encoder module's availability. When present, the interleaving block provides a processing benefit while also boosting the process's resilience by providing temporal variation. By applying a known random sequence, the whitening module scrambles the bit stream into fake white noise and enhances receiver synchronization performance. To mitigate the consequences of multi-path transmission, a spreading module is required to disseminate the cyclic prefix-inserted information modulation. This feature allows the conversion of frames from the time to the frequency domain to be adjusted. Subsequently, the sync insertion module inserts the needed synchronization pattern for receiving processes. The pulse shaping feature of the RRC functions as a digital filter, thereby lowering radiation that is sent at frequencies beyond the transmission radio-frequency band. Attached modules are necessary during the receiving process to coarsen the estimate and correct time offsets. To establish the payload start location, it is essential to determine the burst's start, the estimation and correction of small frequency errors, channel estimation and equalization, and timing recognition.

Weightless, like many other systems, communicates between protocol levels using channels. Based on their function, control, logical, transport, and physical channels are the three categories of channels. Three physical channels comprise the physical layer (PHY), which is responsible for baseband data exchange. They are the uplink, downlink, and uplink contested access channels. Uplink channel communication is used to communicate between an access point and an ED. Data are transmitted by the BS to a single or several EDs through the downlink channel. Additionally, data are sent from EDs to the BS via the uplink contested access physical channel. End devices compete for capacity on this channel, and a large number of EDs are permitted to communicate concurrently.

Of the link layer (LL) and the baseband (BB) sublayer, the BB sublayer is in charge of supplying data to the transport channels. This is accomplished through linking the transport channel to the PHY channel. BB is responsible for a multitude of functions, including identifying organized allocations within a structural framework and transmitting frames in both the directions of downlink and uplink across suitable physical channels. Additionally, BB manages the contested access (CA) operation via the physical channel for contended access uplink. Between BB and LL, a variety of transport mechanisms are used, depending on the sort of addressing information used. The LL establishes connections between logical channels and BB transport channels. Additionally, LL is responsible for the regulation of retransmissions, logical channel reliability, fragmentation, and re-assembly of data. These steps may be followed or disregarded. The LL is responsible for multiplexing and de-multiplexing transport channels into user data channels or logical control. As a result, LL establishes logical channels between ED and BS or BSN using either an unrecognized and defective packet stream or a recognized and dependable packet stream.

The Radio Resource Manager (RRM) is responsible for controlling communication between an ED and its BS via the control channel, while the LL provides adequate security. During communication, RRM makes use of control channels to regulate and control the communications that communicate between the ED and the BS. Additionally, the RRM maintains the ED-BS link via a downlink-only control message stream delivered by the BSN. The Weightless protocol stack is depicted in Figure 4.6.

User channels are used to communicate between an ED and its BS. They provide separate channels for multicast and unicast data transmissions, as well as acknowledgment and interrupt data. The multicast data channel between the ED and the BS is a one-way uplink-only channel, whereas the unicast data channel is bidirectional. Weightless-W is a two-way channel that operates in the frequency of 470–790 MHz TVWS frequency band. The data transfer rate ranges from 1 to 10 Mbps, while the battery span is estimated to be between 3 and 5 years, based on utility. This technique uses a 128-bit AES encryption and star topology for packets.

A packet can include up to 10 bytes of data, although end-to-end encryption can be performed in other ways, such as through the network core. The manufacturer describes an undisclosed FEC algorithm as the basis for the error-correcting technology. Channeling is achieved by utilizing 16–24 channels with a bandwidth of 5 MHz, based on the frequency of usage. The channels are modulated in an adaptive manner.

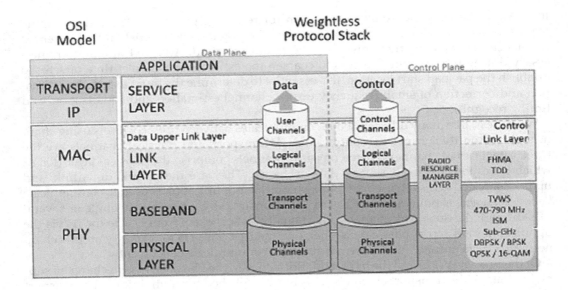

FIGURE 4.6
Weightless protocol stack.

Depending on the application, they may attain high speeds over small distances. As the distance between the receiver and transmitter increases, modulation can begin with DBPSK and move to BPSK. Lower speeds, starting at 1 Kbps, use 16-QAM or QPSK (quadrature phase-shift keying), with a maximum transmission rate of 10 Mbps over short distances [32–35].

4.4.8 Wireless Fidelity (Wi-Fi)

It offers Internet-accessed wireless communication through radio frequencies. Network authorization control, information security, and network planning are all possible attack vectors. As explained below [36], Wi-Fi employs six unique security modes: overall security of Wi-Fi network information; Protected Access via WPA-Wi-Fi, Equivalent Privacy through WEP-Wired, and WPA2 encryption; MAC address filtering; Protocol filtering, SSID (shield service set identifier) broadcast information; and assigning IP address.

Additionally, Wi-Fi is susceptible to the following attacks: (a) evil twin attack, which is an anonymizing hack that misleads users toward linking to a fake Wi-Fi access point. (b) Jamming signals: A hacker can interrupt a Wi-Fi network by blocking the signal, which is also known as "producing noise." (c) Misconfiguration attacks: If a gateway is configured with default credentials and encryption methods, cybercriminals acquire network access. (d) Honeypot attack: An attacker creates bogus gateways with the same SSID as a public Wi-Fi AP, luring users into a trap. (e) Unauthorized/ad hoc connection attacks: A virus or spyware, for example, can be used to activate an ad hoc connection or an employee may already be using one. Ad hoc systems are susceptible to assault because they lack greater encryption. (f) Wireless eavesdropping: Information is obtained from the network by illegally accessing transmitted data.

4.4.9 Constrained Application Protocol (CoAP)

It is an application-layer synchronous request/response protocol designed by the IETF for devices with limited resources. It was designed to be interoperable with HTTP by utilizing a subset of HTTP methods [37].

To keep the implementation as light as possible, CoAP makes use of UDP. It implements resource-oriented interactions in a client-server paradigm by utilizing HTTP commands. To manage resources, UDP-based application layer protocols were developed to eliminate TCP and minimize bandwidth needs [38]. Similarly, unlike TCP, which by design is unicast-oriented, CoAP supports both unicast and multicast.

CoAP included its own algorithms for achieving reliability while running on the unreliable UDP protocol. Each header packet consists of two bits indicating the kind of message and the QoS level that is required. There are four distinct sorts of messages:

1. **Confirmable:** The communication which contains a request for action (ACK). If the response takes additional computation time, it might be supplied synchronously (inside the acknowledgement) or asynchronously (via a separate message).
2. **Unconfirmable:** An unconfirmed transmission.
3. **Acknowledgement:** This indicates that a transmission has been received.
4. **Reset:** Indicates the receipt of a message, but its inability to be handled.

Additionally, for verifiable notifications, there is a simple stop-and-wait resend feature as well as a 16-bit header field called "Message ID" that is unique and used to detect CoAP packet duplication.

CoAPCHTTP Mapping allows CoAP clients to access HTTP servers' resources via a reverse proxy [39]. Despite being designed for IoT and M2M communications, CoAP lacks any in-built security measures. Datagram Transport Layer Security (DTLS) should be used to safeguard CoAP transactions. TLS for TCP is like DTLS, which works on top of UDP. There are options for identification, information integrity and secrecy, key management automation, and cryptographic approaches. It is true that DTLS can protect UDP transmissions, but it wasn't made for the IoT, so its usefulness isn't clear. The first thing to note about DTLS is that it doesn't support multicast. This is a big advantage for CoAP over other application layer protocols. Network traffic, computational resources, and battery life of mobile devices, which are important parts of the IoT, are all affected by DTLS handshakes. Although CoAP is HTTP-compatible due to its IoT design, CoAP over DTLS may cause further difficulty for HTTP servers due to its different packet layout. In the literature, there are other methods for securing CoAP, including ones that are actively being researched [40].

4.4.10 Message Queuing Telemetry Transport (MQTT)

It is a low-power M2M communication protocol established by IBM. It is a publish/subscribe asynchronous protocol relay on TCP. Due to the fact that clients are not required to seek updates, publish/subscribe protocols are more suited to IoT requirements than request/response protocols. As a result, network traffic is reduced and computational resources are required less frequently.

In MQTT, topics are stored in a broker (server). Each client might be a publisher, sending data to the dealer about a particular subject, or a recipient, receiving instant notifications whenever a topic to which he is subscribed receives an update. Because the MQTT protocol is optimized for bandwidth and battery life, Facebook Messenger now uses it.

MQTT ensures reliability by offering three different levels of QoS.

1. **Fire and Forget:** Once alone information is transmitted and is not to be recognized.
2. **At Least Once Delivered:** A message must be transmitted at least once, and it must be recognized.
3. **Delivered Exactly Once:** It is a four-way handshake system utilized to check that the information is sent exactly once.

It is an application layer protocol based on TCP with minimal overhead. Furthermore, for IoT applications, messages must be responded to using the publish/subscribe architecture, for example CoAP request/response. This results in a reduction in network and width, as well as a reduction in message processing, extending the battery-powered device's lifetime.

MQTT brokers may demand username/password verification to maintain security, which is handled via TLS/SSL Secure Socket Layer, the same safety control standard that safeguards HTTP transactions throughout the web.

When comparing MQTT to the previously stated CoAP, it is noticeable that the packet loss is greater while using CoAP due to the lack of TCP retransmission facilities. Based on a recent study [44], MQTT offers fewer delays compared to CoAP when minor packet losses occur. Confirming the accuracy of COAP creates fewer traffic issues. Moreover, based on the network conditions, the results may vary. Meanwhile, the communication's quality of service (QoS) has an effect on packet loss and delay. When the QoS level is raised, both protocols see less packet loss and more delay.

4.4.11 Advanced Message Queuing Protocol (AMQP)

It is a publish/subscribe architecture that is based on a scalable and reliable messaging queue. OASIS has standardized it. Nowadays, AMQP is widely utilized as a part of commercial and business platforms. Due to the publish/subscribe approach, this protocol is extremely scalable.

AMQP enables the communication between devices that support a variety of languages to be heterogeneous and interoperable. Messages can be exchanged between AMQP-enabled applications. The AMQP places a premium on knowledge of a set of message specifications to ascertain the safety, reliability, and system outcome.

In an IoT environment, the AMQP is used to share and convey information. To ensure reliability, AMQP offers three distinct information delivery assurances: exactly once, at least once, and at most once. Additionally, this standard includes a TCP transportation layer for increased reliability. The publish/subscribe mechanism in AMQP is composed of a couple of components: an exchange and a message queue. The exchange queue is in charge of dispatching messages to the correct queue position. Messages are queued until the intended recipient receives them. There is a specialized method for exchanging messages between exchange components and message queues that include a set of rules. AMQP Evaluation: This protocol has a high degree of extensibility and interoperability across a wide range of platforms and environments; it also excels in industrial environments. However, it is incompatible with constrained environments, is not developed for real-time implementations, and lacks an automated discovery method.

4.4.12 Data Distribution Service (DDS)

It is a widely used protocol in today's IoMT environment, and it is a field of research that is rapidly expanding. The Object Management Group (OMG) developed DDS to encourage

IoMT applications as well as M2M communication. It uses a publish/subscribe architecture. The ability to achieve QoS and dependability is one of the reasons this protocol is well suited for M2M and IoMT applications. To ensure reliability, this protocol specifies numerous service quality requirements. This protocol adheres to security, durability, priority, and other standards.

The DDS protocol implements a number of QoS criteria based on the model of publishing/subscription, which requires a robust discovery model to assist subscribers in finding publishers, as this is the key predictor of the protocol's performance and reliability. Between the publisher and the subscriber, DDS sends data as a topic. The Data Readers of subscribers and the Data Writers of publishers generate data on the subject.

4.4.13 Extensible Messaging and Presence Protocol (XMPP)

The XMPP was standardized over a decade ago by the IETF, establishing it as a tried-and-true standard that is widely used on the Internet. Furthermore, as an elderly protocol, it falls short of meeting the requirements of some emerging data applications. As a result of this lack of widespread adoption, Google announced last year that it would stop supporting the XMPP standard. The XMPP, on the other hand, has recently regained popularity as an IoT-friendly communication protocol.

It is a messaging system based on TCP that allows both asynchronous and synchronous messaging. Due to its near-real-time nature, it has tiny message footprints and less text latency.

TLS/SSL protection has been developed into the XMPP. However, it lacks QoS capabilities, rendering it unsuitable for M2M communications. TCP's dependability is ensured solely by its inherited mechanisms.

In comparison with CoAP's synchronous model, the XMPP supports an asynchronous architecture, which is more suited to the IoT. Additionally, it is a well-established protocol with widespread Internet support, which gives it a significant advantage over the relatively new MQTT. However, XMPP messages are written in XML (Extensible Markup Language), which adds unnecessary tag overhead and necessitates XML parsing, which requires additional computational capability and thus increases power consumption.

4.4.14 WebSocket

WebSocket was created as part of the HTML initiative to streamline TCP communication. It is neither a publish/subscribe nor a request/response protocol. A handshake is created between the client and the server in order to start a WebSocket session. The session can be enabled by clients as well as servers to communicate through full-duplex mode asynchronously. When no longer required, a session can be terminated either from the client side or from the server side. This protocol was designed to minimize the overhead associated with Internet/web connections while maintaining full-duplex communication in real time.

The WAMP is a sub-protocol of WebSocket that enables publish/subscribe messaging systems. WebSocket, which is based on the robust TCP protocol, lacks any built-in reliability measures. If necessary, a WebSocket over TLS/SSL can be used to secure sessions. WebSocket transmissions have a session overhead of only two bytes. As with the preceding protocols, WebSocket is not intended for devices with limited resources, and its client-server architecture is incompatible with IoT applications. However, it is intended for real-time communication, is protected, reduces overhead, and can enable better-messaging systems when combined with WAMP. As a result, this could be used in conjunction with any other TCP.

4.5 Conclusions

IoMT is an emerging technology that utilizes different communication protocols for different applications, each and every communication protocol will have its own features, and it's wise to choose the appropriate protocol that supports the system with a secure network. It can be preferred by considering its own merits and demerits. Since the hardware requirements and size of the device and the place where the device needs to use will decide its own communication protocol and the technology is still growing. This chapter summarized the various communication protocols used in IoMT environments.

References

[1] Choudhary, G., Jain, A.K. Internet of things: A survey on architecture, technologies, protocols and challenges. In: *IEEE International Conference on Recent Advances and Innovations in Engineering (ICRAIE-2016). Jaipur, India*, 23–25 December, 2016.

[2] Belkeziz, R., Jaris, Z. *A Survey on Internet of Things Coordination.* IEEE, 2016, 978-1-5090-4926-4/16/$31.00.

[3] Lin, J., Yu, W., Zhang, N., Yang, X., Zhang, H., Zhao, W. *A Survey on Internet of Things: Architecture, Enabling Technologies, Security and Privacy, and Applications.* IEEE, 2016, pp. 2327–4662.

[4] Al-Fuqaha, A., Guizani, M., Mohammadi, M., Aledhari, M., Ayyash, M. *Internet of Things: A Survey on Enabling Technologies, Protocols and Applications.* pp. 1553–877X, 2015.

[5] Atzori, L., Iera, A., Morabito, G. The internet of things: A survey. *Comput. Networks.* 54, 15, 2787–2805, 2010.

[6] Laranjo, I., J. Macedo, A. Santos. Internet of things for medication control: Service implementation and testing. *Procedia Technol.* 5, 777–786, 2012.

[7] Santos, A., J. Macedo, A. Costa, M. J. Nicolau, Internet of things and smart objects for M-health monitoring and control. *Procedia Technol.* 16, 1351–1360, 2014.

[8] Bhaskar, K.B., A. Prasanth, P. Saranya, An energy-efficient blockchain approach for secure communication in IoT-enabled electric vehicles. *Int. J. Commun. Syst.* 35, 1–25, 2022.

[9] Lavanya, S., A. Prasanth, S. Jayachitra, A Tuned classification approach for efficient heterogeneous fault diagnosis in IoT-enabled WSN applications. *Measurement.* 183, 1–22, 2021.

[10] Ramasamy, S. M. D., K. Periasamy, L. Krishnasamy, R. K. Dhanaraj, Multi-disease classification model using Strassen's half of threshold (SHoT) training algorithm in healthcare sector, *IEEE Access.* 9, 112624–112636, 2021.

[11] Xu, B. et al., Ubiquitous data accessing method in IoT-based information system for emergency medical services, *IEEE Transac. Ind. Inform.* 10, 2, 1578–1586, 2014.

[12] Santos, D. F. S., H. O. Almeida, A. Perkusich, A personal connected health system for the internet of things based on the constrained application protocol, *Comput. Electr. Eng.* 44, 122–136, 2015.

[13] Abinaya, V. K., Swathika, Ontology based public healthcare system in internet of things (IoT), *Procedia Comput. Sci.* 50, 99–102, 2015.

[14] Gomez, J., B. Oviedo, E. Zhuma, Patient monitoring system based on internet of things, *Procedia Comput. Sci.* 83, 90–97, 2016.

[15] Rahmani, A. M., et al., Exploiting smart e-Health gateways at the edge of healthcare Internet-of-things: A fog computing approach, *Futur. Gener. Comput. Syst.* 78, 641–658, 2016.

[16] Moosavi, S. R., et al., End-to-end security scheme for mobility enabled healthcare Internet of Things, *Futur. Gener. Comput. Syst.* 64, 108–124, 2016.

[17] John, J. T., S. J. Ramson, Energy-aware duty cycle scheduling for efficient data collection in wireless sensor networks, *IJARCET.* 2, 2013.

[18] Ramson, S., D. J. Moni, A case study on different wireless networking technologies for remote health care, *Intelligent Decision Technol.* 10, 4, 353–364, 2016.

[19] Al-Fuqaha, A., M. Guizani, M. Mohammadi, M. Aledhari, M. Ayyash, Internet of things: A survey on enabling technologies, protocols, and applications, *IEEE Commun. Surv. Tutorials.* 17, 4, 2015.

[20] Mike Ryan, et al. Bluetooth: With low energy comes low security, *WOOT.* 13, 4–4, 2013.

[21] Prasanth, A., S. Pavalarajan, Zone-based sink mobility in wireless sensor networks, *Sensor Rev.* 39, 6, 874–880, 2019.

[22] Roy, S., Jandhyala, V., Smith, J.R., Wetherall, D.J., Otis, B.P., Chakraborty, R., Buettner, M., Yeager, D.J., Ko, Y.C., Sample, A.P. RFID: From supply chains to sensor net, *Proc. IEEE.* 98, 1583–1592, 2010.

[23] Dhiviya, S., Malathy, S., Kumar, D. R. Internet of things (IoT) elements, trends and applications, *J. Comput. Theoret. Nanosci.* 15(5), 1639–1643, 2018.

[24] Coskun, V., Ozdenizci, B., Ok, K. A survey on near field communication (NFC) technology. *Wirel. Pers. Commun.* 71, 2259–2294, 2013.

[25] A. Prasanth, S. Jayachitra. A novel multi-objective optimization strategy for enhancing quality of service in IoT-enabled WSN applications, *Peer-to-Peer Netw. Appl.* 13, 1905–1920, 2020.

[26] Want, R. Near field communication, *IEEE Pervasive Comput.* 10, 4–7, 2011.

[27] Nobre, M., Silva, I., Guedes, L.A. Routing and scheduling algorithms for wireless HART networks: A survey, *Sensors.* 15, 9703–9740, 2015.

[28] Iordache, V., Gheorghiu, R.A., Minea, M. Analysis of interferences in data transmission for wireless communications implemented in vehicular environments. In *Proceedings of the Federated Conference on Computer Science and Information Systems (FedCSIS), Prague, Czech Republic,* pp. 849–852, 2017.

[29] Song, J., Chen, D., Nixon, M., Lucas, M., Pratt, W., Han, S., Mok, A. WirelessHART: Applying wireless technology in real-time industrial process control. In *Proceedings of the IEEE Real-Time and Embedded Technology and Applications Symposium (RTAS), St. Louis, MO, USA,* pp. 377–386, 2008.

[30] Fuller, J. D., and Ramsey, B. W. Rogue z-wave controllers: A persistent attack channel. *2015 IEEE 40th Local Computer Networks Conference Workshops (LCN Workshops).* 734–741.

[31] Weightless Special Interest Group (SIG). Weightless—Setting the Standard for IoT. Available online: http://www.weightless.org/.

[32] Adelantado, F., Vilajosana, X., Tuset-Peiro, P., Martinez, B., Melia-Segui, J., Watteyne, T. Understanding the limits of LoRaWAN, *IEEE Commun. Mag.* 55, 34–40, 2017.

[33] Weightless-SIG. Weightless specification. Available online: http://www.weightless.org/about/weightlessspecification.

[34] Webb, W. *Understanding Weightless: Technology, Equipment, and Network Deployment for M2M Communications in White Space.* Cambridge University Press: Cambridge, 2012.

[35] *Weightless Special Interest Group (SIG) Weightless-P System Specification.* Available online: http://www.weightless.org.

[36] Grabovica, M., D. Pezer, S. Popić and V. Knezević, Provided security measures of enabling technologies in Internet of Things (IoT): A survey, *Zooming Innovation in Consumer Electronics International Conference (ZINC),* 2016.

[37] Castellani, A.P., M. Gheda, N. Bui, M. Rossi, M. Zorzi, *Web Services for the Internet of Things through CoAP and EXI, IEEE International Conference on Communications Workshops,* ICC, pp. 1–6, 2011.

[38] Sye Loong, K., S.S. Kumar, H. Tschofenig, Securing the Internet of things: A standardization perspective, *Internet Things J. IEEE.* 1, 3, 265–275, 2014.

[39] Maria Rita, P., N. Accettura, X. Vilajosana, T. Watteyne, L.A. Grieco, G. Boggia, M. Dohler, Standardized protocol stack for the Internet of (important) things, *Commun. Surv. Tutorials IEEE.* 15, 3, 1389–1406, 2013.

[40] T.A. Alghamdi, A. Lasebae, M. Aiash, Security analysis of the constrained application protocol in the internet of things, *Second International Conference on Future Generation Communication Technology (FGCT),* pp. 163–168, 2013.

5

Security and Privacy of Biomedical Data in IoMT

Ashish Kumbhare and Piyush Kumar Thakur
The ICFAI University

CONTENTS

DOI: 10.1201/9781003256243-5

5.1 Introduction

Nowadays, the healthcare sector is one of the industries that have been a leader in implementing new technologies that provide ubiquitous and real-time services. Under the banner of IoT, a significant percentage of entities, including individuals, machines, and objects, are interconnected into the information environment at any time and from any location. The evolution of IoT is revolutionizing the healthcare business and bringing in the Internet of medical things (IoMT), in which medical equipment is connected in a worldwide network to which anyone, anywhere, and at any time may connect. The integration of IoT technique medical equipment into the IoT gave rise to the IoMT.

The landscape of IoMT system has taken a significant impact in terms of health services, pushing millions of people across the world to earn a healthy living [1]. In this setting, healthcare services have evolved into consumer, precise, and customized services, including a private medical professional available all the time. The advent of the IoMT has enhanced remote patient monitoring. It links patients with their specialists and allows for the safe transmission of patient data, reducing the frequency of needless medical appointments and the load on healthcare organizations.

By allowing patients to communicate health-relevant information to practitioners, IoMT offers the potential to provide more efficient treatment, fewer mistakes, and cheaper healthcare expenditures [2]. This is now important because of the worldwide pandemic, COVID-19 that is reducing in-person medical appointments and thereby avoiding disease spread. IoMT has enhanced remote patient monitoring that assists in the observation of patients' vital signs with specialists being contacted instantly when necessary. Those who live in remote places can use smart devices to share activity tracker data with a remote health professional and receive a medically informed recommendation [3]. The IoMT insurance firms can evaluate patient data more quickly and process claims more efficiently and precisely, thanks to the Internet of things. Because it enhances patient care quality, IoMT benefits all stakeholders, including insurance companies and pharmaceutical businesses (Figure 5.1).

IoMT is a subset of the Internet of things that includes smart devices such as wearable and medical or vital monitors that are designed to track people's health. Smart electronics are included in several of these wearables with the ability to measure and broadcast a user's physiological parameters in real time. The "things" in the IoMT can refer to a wide range of devices that provide a pre-programmed level of fluids into a patient, such as infusion pumps and cardiac monitoring implants.

Numerous additional devices, such as pacemakers, insulin pumps, and cochlear implants are also available. These gadgets capture and transfer data to healthcare practitioners over the Internet. It permits healthcare providers to monitor a patient's health from afar. It also helps them to respond quickly to problems as they arise, rather than waiting for patients to visit the doctor in person. IoMT also refers to the linking of software programs that collect medical data from Internet computer networks and send them to hospital IT systems.

IoMT devices, in particular, which make up the essential underlying pieces of the IoMT edge network (e.g., implantable sensors and medical wearable), are exposed to a wide range of security assaults, creating a significant risk to patients' safety and privacy. The simple truth that security is a critical component that is largely dependent on the dependability of the healthcare devices involved, innovative security measures to preserve the safety of the IoMT network are urgently needed for the effective implementation of IoMT

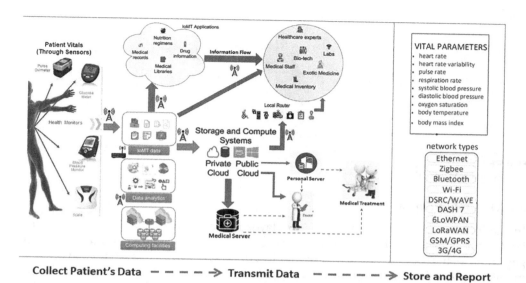

FIGURE 5.1
Typical IoMT system's components and the stakeholders.

technology into the wider healthcare system. To that aim, the first step is to gain a thorough awareness of actual and future risks to the IoMT system and properly categorize them. Because IoMT equipment has abilities and technological features comparable to IoT devices, known threats targeting IoT networks might likewise be regarded as possible risks to the IoMT network. As a result, the authors conducted a comprehensive research on the current and anticipated security risks to the IoMT network environment and classified them based on the major security goals that these threats aim to achieve [4].

5.1.1 IoMT Architecture

IoMT allows IoT communication protocols to be combined with medical systems and equipment to support remote, real-time patient treatment and monitoring. Smart hospitals, for example, use IoMT to "offer automated and optimized procedures that are based on (ICT) Information and Communication Technologies environment of interlinked assets, especially based on IoT, to improve and introduce new capabilities and existing patient care procedures." Because most communication protocols haven't been explicitly established for the demands of linked medical devices, there's a need to analyze the security of accessible IoT connectivity in the context of medical equipment.

We employ a categorization based on IoT communication protocols' three levels, application layers, network layer, and the perception layer to complete this evaluation. Each layer, like the OSI (Open Systems Interconnection) layer, effectively incorporates distinct sorts of protocols and methods for information sharing. The IoMT architecture aids in a better understanding of the system's multiple layers' composition. Several layers have been proposed in several articles, with various nomenclatures for these layers. The researchers have yet to settle on a single or generic architecture for IoMT healthcare system (IHS). The advancement of IoMT, application needs, and the difficulty of IoMT considering privacy and security are among the key reasons for the various IoMT designs (Figure 5.2).

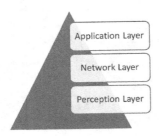

FIGURE 5.2
The layered IoMT architecture.

5.1.1.1 Application Layer

In the IoMT architecture, the application layer is the topmost layer. It provides customized services based on the demands of the user. Its major task is to bridge the significant gap that exists across consumers and applications. At the device level, this layer is implemented using a specific program. Application layer protocols specific to IoMT exist, just as they do for a browser implementing FTP, HTTP, HTTPS, and SMTP on a PC. The HTTP is commonly used as the application layer on the Internet. HTTP, on the contrary, is inappropriate for resource-constrained environments such as IoMT since it is exceedingly heavy; it incurs a large processing overhead [5].

For IoT contexts, a variety of alternative protocols have been established. Message Queuing Telemetry Transport, Constrained Application Protocol, Secure Message Queue Telemetry Transport, XMPP, and WebSocket are some of the most popular IoMT application layer protocols. The privacy of data is a serious problem for this layer, as diverse devices communicate via multiple IoMT protocols. Different devices are communicating a vast amount of data, which must be protected. It is responsible for providing all necessary data in response to user queries, with privacy being a key component. There are also other data privacy techniques used, which include Secure Sockets Layer, Transport Layer Security, and domain name system. Lots of devices can be found in this layer such as monitoring system, fitness/health system, remote diagnosis system, telemedicine, medical e-record, and tracking/locator system.

5.1.1.2 Network Layer

Packet routing and network addressing from source to destination nodes are handled by the IoMT network layer. It collects and sends data from all devices, as well as connects them to other smart items, servers, and network devices. A non-wired or wired transmission medium can be used. Bluetooth, Ethernet, 3G, RFID, wireless LAN, and NFC are the technologies utilized at this layer for IoMT. Ethernet links IoMT devices that are stationary or fixed. Because IoMT devices might be stationary and require a stable power supply, Wi-Fi systems are used to link the gateway to the end-user.

Many low-power IoMT devices communicate to end-users and other nodes using radio spectrums. Some healthcare gadgets in healthcare center use Wi-Fi or a low-power personal area network (WLAN) to connect to other devices [6]. Wearables commonly use Bluetooth for communication. Bluetooth Low-Energy (BLE) and Zigbee were created particularly for low-power devices. It only transmits little quantities of information and isn't designed to handle massive datasets. The fact that Zigbee can accommodate large number of nodes is its most outstanding feature. Because of its low-power characteristics, it is ideal for IoMT devices.

5.1.1.3 Perception Layer

This layer's primary job is to gather data (temperature, heart rate, pressure, and so on) and then transmit them to the network layer. The sensors that detect and collect environmental data are placed in this physical layer. The perception layer is the typical architecture of the IoMT. It identifies other intelligent entities in the surroundings and translates their physical features into electrical impulses.

Interconnections among medical devices can be safe and constantly available because of the extremely sensitive nature of information processed in IoMT. Furthermore, the medical records shared on the hospital's network must maintain confidentiality, data integrity, and availability. Different operational and interoperability difficulties must also be considered [7]. For example, the variety of medical assets employed all around the healthcare ecosystem necessitate the integration of many technologies into a single IoMT ecosystem.

5.2 Background

5.2.1 Privacy and Security Needs for IHS

IHS has more stringent privacy and security requirements than normal IoT-based infrastructures. Many additional securities needed for IHS exist, such as device localization, which can help to secure the systems' privacy and security. Each level of the IHS has distinct functionality, which means that each level has distinct privacy and security needs. As a result, each level's criteria are examined and described separately. Furthermore, in the subject of the GDPR and HIPAA, the privacy and security standards at the data level are examined.

5.2.1.1 Requirements at Data Level

a. **Confidentiality:** Patient's medical data must be collected and stored under ethical and legal privacy rules, which limit access to only approved personnel. Appropriate steps must be taken to safeguard the privacy of health information linked with specific patients to avoid data breaches. The necessity of such protection cannot be emphasized, since data obtained by cyber-theft may be sold on black markets, putting patients at risk of not just privacy violations, but also reputational and financial damage.

b. **Integrity:** The goal of the information integrity necessity for IHS is to verify that information reception at the desired location has not been tampered with in any manner in transition. Intruders might access and alter patient records by using the wireless network's communication feature; in life-threatening situations, this might have massive consequences. The ability to identify any unauthorized data distortions or modifications is crucial for ensuring that data haven't been compromised. As a result, proper data integrity safeguards must be developed to prevent malicious attempts from altering sent data. Furthermore, the integrity of data saved on medical servers must be guaranteed, which implies that the data cannot be tampered with. Medical service providers must take reasonable steps to ensure that patient data are correct and up to date. Inaccurate personal data must also be destroyed or corrected as quickly as feasible.

c. **Availability:** Data and services must be available to the appropriate users when they are needed. If DoS cyber-attacks are conducted, medical servers and devices data would be rendered inaccessible. Any unavailable data or services might result in life-threatening situations, such as the inability to deliver timely alerts in the event of a heart attack. As a result, to account for the possibility of loss of data, healthcare apps should be constantly available to ensure that data are available to users and emergency services. Medical service providers may be able to regain access and availability to private data in a timely way, which includes implementing preventative security precautions and countermeasures to denial of service (DoS) attacks.

5.2.1.2 Requirements at Sensor Level

Because of the limited processing capabilities and power constraints of medical equipment and sensors, in terms of privacy and security, the sensor layer of the three-tier IHS poses the most concerns (Figure 5.3).

a. **Tamper-Proof Hardware:** Physical theft of IoMT equipment, particularly ambient sensors, might expose security information to intruders. Furthermore, intruders may reprogram and re-deploy stolen devices to the system, eavesdropping in on conversations without even being discovered. As a consequence, medical device theft has become a major security issue in IHSs. The systems' medical equipment must always include tamper-proof ICs, which prevent third parties from accessing codes recorded on the devices after they've been implemented [8].

b. **Localization:** Patient placement and on-body sensor positioning are the two forms of sensor localization that researchers are focusing on. For programs such as hobby recognition, the former type of sensor localization is utilized to determine

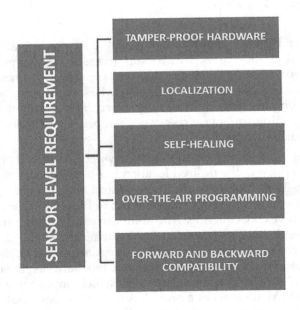

FIGURE 5.3
Privacy and security requirements at sensor level.

whether or not the sensors are in accurate physical positions. Sensor localization is a technique for locating a sensor inside a room or a patient wearing a sensor within a structure. Furthermore, medical devices may move into or out of network coverage often due to the nature of IHS. As a result, comprehensive intrusion detection systems (IDS) are needed if the sensors leave and rejoin at random intervals.

c. **Self-Healing:** Self-healing is critical for IoMT systems, as stated in autonomic computing, since IoMT devices must continue to function after network assaults. An IoMT system is expected to identify and diagnose assaults and apply appropriate security methods with little human interaction to accomplish self-healing. Self-healing solutions should be as low as possible in terms of network communication costs and processing difficulty for medical and healthcare equipment. However, because different forms of network assaults need distinct detection and recovery mechanisms, network managers must select which autonomic security strategies should be used.

d. **Over-the-Air Programming:** OTA is a means of updating an IoT network with a sensor, but it poses security concerns regarding rogue sensor nodes that are looking for updates and introducing bogus identities into the network. OTA can be utilized in conjunction with a self-healing procedure to keep network security policies up to date. To successfully utilize OTA, security precautions must be put in place to prevent attackers from exploiting OTA updates. SEDA, for example, is a safe OTA programming technique developed for dispersed networks such as IoMT systems.

e. **Forward and Backward Compatibility:** This is especially required in healthcare applications, where defective sensors must be replaced as quickly as feasible. Medical sensors will not be able to receive future transmissions if they are sent after the sensor leaves the network; this is known as forwarding compatibility. Backward compatibility, on the other hand, prevents messages from being examined by a sensor that has joined the network. Compatibility concerns may be resolved by using OTA programming to distribute the most recent software update as soon as possible.

5.2.1.3 Requirements at Personal Server Level

Before sending to medical servers, patients' data are usually held and collected on personal servers in IoMT systems; therefore, ensuring that the records are securely handled even as on the personal servers is essential. To hold privacy and security on the private server level, two types of authentication methods have to be used: person authentication and tool authentication.

a. **Person Authentication:** Effective user authentication techniques are crucial for the reason that data stored on private servers, whether permanent or temporary, must only be accessible with the aid of using patients and medical workers, along with caretakers. Personal servers in IHS should offer urgent availability of data in critical situations, such as experiencing a stroke. Biometrics is a frequent approach for person validation at the personal server level, and they're especially valuable in IHS because the majority of biometrics may be obtained simply by implanting or wearing medical equipment in the human body [9].

b. **Tool Authentication:** Before receiving data from medical devices and sensors, the personal server must undertake authentication. For data security and integrity, secure/encrypted communications should be possible using a device/tool authentication mechanism. Because inaccurate information from dangerous devices concerning patients' physiological parameters might have major serious effects for clinical diagnosis and treatment choices, device/tool authentication must be included in any IHS. Device/tool authentication is bidirectional among personal servers and tools, but because personal servers frequently have greater power and computational capability than medical sensors and devices, the majority of the computation should be done on them.

5.2.1.4 Requirements at Medical Server Level

Only authorized items and individuals have direct exposure to patients' medical information on the medical server level, and data must be fully protected while stored in databases. Privacy and security problems with medical servers containing electronic medical records (EMRs) are developing as more paper-based medical information is transformed into EMRs. As a result, suitable security features for IHS need to be in areas on the clinical server level [10].

a. **Restriction of Access:** Only approved workers and devices should have direct exposure to clinical servers, according to effective access control procedures. Because obtaining a patient's consent or permission whenever a data access request is made is extremely difficult, medical server service providers must provide patients with selective data access permission, allowing patients to identify whose information may be disclosed before consent and which service providers have accessibility to it. ABE (attribute-based encryption), a kind of asymmetric cryptography in which private keys are created based on characteristics, is a popular choice for access control. In ABE solutions, access trees may be built selectively using a collection of characteristics such that only those qualities that satisfy the tree are permitted access to the protected data. Medical servers must also be able to quickly update access control policies. For medical servers, policy updates can be repetitive. For example, many cloud protection solutions contain the converting of encryption keys while upgrading access control policy, resulting in data being decrypted and re-encrypted on both personal and medical servers. To minimize or remove processing overheads in cryptography, a scalable and less repetitive policy updating technique should be used. A prominent approach is two-layer over-encryption, which allows policy updates in the surface encryption layer (SEL) while data owners enforce additional encryption in the base encryption layer (BEL). Urgent restriction of access must be provided by healthcare servers, by removing safeguards for security on health records or by authorizing third-party secure access [11].

b. **Key Distribution:** Efficient key distribution methods are required for the construction of secure transactions, with the purpose of implementing and transmitting cryptographic functions/keys to sensor devices. Key distribution and trusted server are two major kinds of key management protocols used in IHS. Although these protocols are acceptable for hierarchical networks, they are insufficient for essential applications such as those in healthcare since a full network failure might immobilize a trusted server in a practical situation. In a symmetric-key cryptosystem, key pre-distribution methods are frequently used to transfer secret keys inside a network before it is completely operational. Because their execution is simple and does not need extremely complicated computing, these sorts of resource-limited sensor networks are popular [12].

c. **Trust Management:** The term "trust" refers to a two-way relationship between two dependable nodes that share data, such as sensors and a network coordinator. It is described as the degree to which a node is safe and trustworthy while interacting with another node. Wireless healthcare applications require distributed collaboration across network nodes to be successful. In this context, trust management systems may be used to identify a node's degree of trust, which is crucial since trust evaluation of a node behavior, such as quality and data delivery, is critical in IHS.

d. **Safeguard to DDoS Assaults:** Wireless medical applications are the target of some of the most common DDoS attacks. Attackers can utilize high-energy signals, such as jamming attacks at the perception layer, to obstruct the wireless network's proper operation. Many solutions, such as evasion protection and competition tactics, have been proposed to defend and self-repair the network against this type of assault, but they are all still in the early phases of development. Because wireless networks are dynamic, extensive research is necessary to create solutions to secure real-time IHS from DoS assaults [13].

5.2.2 IHS Security Schemes

The most up-to-date security techniques for IHS are described in this section. In terms of cryptosystem designs, security analysis, and application, a full explanation of the comparative research is offered. Furthermore, the random number generator (RNG) is a necessary component of cryptosystems, as described, with an emphasis on RNG research suitable for IoMT devices. There's also a look at security measures for implanted IoMT devices, as well as a look at biometric authentication and how it's used in IHS [14].

a. **State-of-the-Art:** Asymmetric and symmetric cryptographic algorithms are the two most popular forms. Asymmetric cryptosystems give higher security protection than symmetric encryption, but they demand much more processing power. Because IoMT devices have limited processing capabilities at the sensor level, any cryptographic approaches suggested for protecting IHS should be lightweight, with little communication channel overhead. Data transfer across medical servers and personal servers indeed is safeguarded with significantly better security measures since data are constantly exchanged via the Internet. Over symmetrical cryptography, public-key cryptography is used inside the bulk of access control, cloud-based authentication, and data storage studies.

Symmetric cryptographic methods, on the other hand, are regularly utilized in research on restricted access and data switch in bidirectional form IoMT sensors because they're inexpensive on those devices with limited resources. In hybrid security approaches, symmetric cryptography methods are commonly used, such as session keys. Eavesdropping, replay, insider, impersonation, and MITM attacks are among the most often employed attacks in security research. Several studies look at their security approaches concerning forwarding security (FS), mutual authentication (MA), contextual privacy (CP), anonymity and traceability (A&T), and the ability to unlink. With the exception of that employed genuine gear in their research, the others relied on simulations on computer. Even if most state-of-the-art safety approaches simulate using publicly accessible RNGs, on-node RNG for IoMT applications is still a work in progress.

To generate random numbers in modern computers, a pseudorandom number generator (PRNG) with a random seed is usually employed. PRNGs are software implementations of deterministic techniques. PRNGs with the equal seed will constantly create the equal random series of protocols, which is better for numbers. Potential attackers can derive the PRNGs if the seed isn't produced from an actual random source. Many actual random number generators based on the unpredictability of physical processes aren't suited for tiny sensors because of the size and power constraints of IoMT devices. The utilization of inertial sensors on IoMT devices is one example of a solution for generating real random numbers [15].

b. **Biometric Authentication:** To authenticate identification, a variety of variables might be employed. Statistics can be components of knowledge, like the user's secret, which are verified components in the intrinsic factors, or possessive factors, like the user's characteristics. Most professional IoMT devices now available for tracking fitness and wellness, notably smart watches, employ alphanumeric passwords rather than biometric identification. Researchers are searching into the usage of biometric intrinsic characteristics that might be unique to the individual in IHS since these elements are thought to be more difficult for attackers to breach than the short passwords frequently employed in smart watches (Figure 5.4).

IHS that uses biometric-based security methods should be able to fulfill the standards. Identification and verification are two common acts performed by biometric-based security systems. The comparison of a sample against all the samples in a database is called

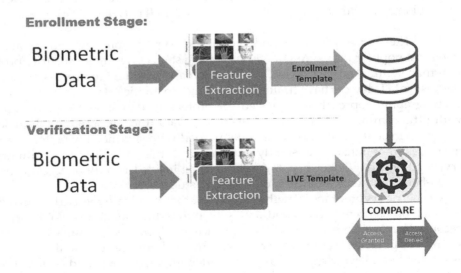

FIGURE 5.4
Biometric authentication systems.

identification, while the comparison of an input dataset between one person's data within the repository is known as verification. Biometric authentication systems are divided into phases: enrollment and matching. During the enrollment stage, participants register their original biometric information in the repository, following which the biometric information is transformed into a characteristic vector and saved in the repository.

A comparable system is used in the matching step. Only if the subject's sample suits the templates or characteristic feature of the authentication process in the database will he or she be validated. Otherwise, the system will deny the log-in attempt. For authentication, a broad diversity of human physical biometric traits can be used. The bulk of physical biometric features, such as fingerprints, palm prints, hand geometry, faces, retina/iris, body odor, vein pattern, ear shape, and DNA, have been utilized in biometric security systems in recent years. Many researchers have turned their focus to multi-biometric fusion to obtain a better degree of security [16].

5.2.3 The Relationship between IoMT Vulnerabilities and Threats

A security assault typically compromises a system by exploiting one or more weaknesses in its components. Table 5.1 keeps track of the relationships between IoMT threats and vulnerabilities.

TABLE 5.1

The Relationship between IoMT Vulnerabilities and Threats

Security Compromised	Stride Threats	Attacks	Striking Vulnerability
Authentication	Spoofing	Spoofing attacks Masquerading attacks MITM attacks	Insecure web interface, unencrypted services, removal of physical storage, insecure cloud interface, insecure network services
Availability	DoS threats	DoS attacks	No account lockout, insecure cloud interface, missing authorization
		Ransomware	Command injection flaw, weak passwords, unencrypted services, insecure cloud interface, insecure network services
Integrity	Tampering	Message modification/replay/fabrication attacks	Unencrypted services
		Physical attacks	Removal of physical storage, command injection flaw
		Traffic analysis attacks	Insecure cloud interface, insecure network services, unencrypted services
Authorization	Elevation of privileges	Elevation of privileges attacks	Unlimited resource allocation
		Malware attacks	Unprotected cloud interfaces, untrusted network services, command injection flaw
Confidentiality	Information disclosure	Social engineering	Removal of physical storage
		Malware attacks	Unprotected cloud interfaces, untrusted network services, command injection flaw
		Eavesdropping attacks	Unprotected cloud interfaces, untrusted network services, unencrypted services, insecure web interface

5.3 Findings

5.3.1 Security Plans for Installable IoMT Devices

Installable IoMT devices are usually implanted into patients' bodies through surgery. As a result, implantable device security schemes have imposed constraints on communication overhead, support for emergencies, power consumption, and attack resilience. Aside from the problems listed above, implantable device security methods must also adhere to strict laws [17] (Figure 5.5).

a. **IMD Protection:** The deployment of a secondary device such as a "proxy" between both the implanted and the output device is the foundation of proxy-based implant security. This technique has the advantage of attempting to increase the safety of currently implanted devices. This is demonstrated by the "IMD-Shield." Noise is introduced into the communication connection between the device and implant to interact with it to achieve "shielding." The implant signal may be decoded at a proxy using the knowledge of the created noise. Only valid communication is routed to and from the implant, according to a security policy. The "IMDGuard" is another proxy-based IMD safety feature that may send keys between the guardian and the IMD.

b. **Distance Bounding:** The wireless transmission length between an external device and implant is restricted by distance bounding, also known as proximity-based access control. Inductive connections are designed to function over small distances and are ideal for charging and programming devices; they lack the data transfer capacity of contemporary technologies. Medical Implant Communication System (MICS) with a greater bandwidth has been chosen by implant manufacturers that

FIGURE 5.5
Security plans for installable IoMT devices.

works in the 402–405 MHz spectral range, with connections from the implant being restricted to a 2 m. Implant information activities could be streamed at 1 m via bedside devices.

c. **Analog Security:** Implants with poor sensor designs are vulnerable to "analog attacks." Sensors are often crucial in a closed loop, such as integrated insulin pumps. Because the sensor signal is basically analog, it may be interfered with, resulting in incorrect sensor data and implant operation. Following appropriate design standards, such as using insulated cables for data transmission, can reduce the disruption of analog signals, which are frequently of modest amplitude, caused by purposeful noise injection.

d. **Power Drawing:** This safety feature was designed to guard against "power drawing" assaults, in which the implant battery is depleted by repeatedly requesting to interface with it. Non-battery sources, also including piezoelectric RF harvesters, must be used to activate all implant transmissions to achieve power communication. While signaling during communication activation improves patient security awareness, radiofrequency energy harvesting may also be used to achieve zero-power communication. The only way for zero-power communication to operate is if the devices are quite near to each other, which restricts its usefulness.

e. **Anomaly Detection:** Anomaly detection can identify resource depletion assaults, which can significantly diminish an IMD's battery power, by studying transaction behaviors among acceptable external devices and IMD. A smartphone analyzes physical layer parameters such as time of arrival, along with behavioral factors such as frequency and range value, of transmissions into and out of IMDs to recognize potentially harmful communications in Medion. Medion has one flaw: It solely safeguards the integrity of IMD; thus, other security techniques should be utilized to safeguard the implanted devices' confidentiality and availability.

5.3.2 Targeted Security and Privacy Aspects in IoMT

Various sorts of cyber-attacks appear to be vulnerable to IoMT privacy and security. Cyber-attacks are classified based on the security features that are targeted. This section will look into security attacks aimed at IoMT data protection considering availability, confidentiality, and integrity. On either hand, we want to look at safety assaults aimed at system security.

5.3.2.1 Breach of Data Confidentiality

Gathering information is required to ensure the confidentiality of IoMT data; because IoMT wireless connections are public and open, patients are subject to interception through privacy intrusions (sniffing) [18]. As a result, there's a good chance that private and personal information may be stolen, hijacked, altered, or leaked. However, several passive assaults can be used to accomplish this (Table 5.2).

a. **Eavesdropping Attacks** are split into two types based on data gathering methods. The first is inactive eavesdropping, which involves scanning Wi-Fi access points to determine whether medical devices are linked to them. Active eavesdropping is the second category. This allows an attacker to monitor inbound and outgoing data while they are in transit, allowing them to capture more information more quickly and easily.

TABLE 5.2

Various Types of Breach of Data Confidentiality with Solutions

Breach of Data Confidentiality	Possible Cause(s)	Solutions
Eavesdropping	Wireless channels are used to transmit messages. A non-encrypted communication channel	Encryption
Interception of data	Wireless communication Unsecured channels	Encryption
Sniffing attacks	Wireless communication Unsecured channels Poor encryption	Encryption
Wiretapping	Wireless communication Unsecured channels	Secure communications Closed communications
Dumpster diving	Inadequate staff training Inadequate awareness	Improved staff training Paperless procedures

b. **Interception of Data:** When a MITM assault is carried out, it results in attacks. An attacker can use this to intercept data and retransmit them later. This allows an adversary to intercept a handshake by eavesdropping on an Address Resolution Protocol (ARP) request and doing it again and again. Then, using this handshake, get the cryptographic keys and get unauthorized entry to clinical data.

c. **Sniffing Attacks**, also known as packet capturing attacks, involve capturing unencrypted medical data packets and disclosing their contents, such as patients' health conditions and credentials. A network monitoring software program such as Wireshark is a good example.

d. **Hacking** medical communications and tele-healthcare equipment to intercept transmitting healthcare data in real time is an example of a wiretapping attack.

e. **Dumpster Diving Attacks** involve searching the dumpster and obtaining all medical information, including patient records, medical prescriptions, employee identities, and other papers and files deposited in the garbage. This is the primary reason why so much data and file records are turning digital.

5.3.2.2 Attacks Employing Social Engineering (SE)

It is a methodology that is used to mislead people by luring them into giving out information. To engage in a cyber-attack later, this comprises passwords, names, IDs, and private information. It appears that luring individuals can be accomplished more readily by depending on human feelings rather than targeting a system's weakness [19]. As a result, the intruder preys on a person's curiosity or lust for adult photographs to get access to health systems or records (phishing), for example (Table 5.3).

a. **Person-to-Person Attacks:** Reverse social engineering is another term for a person-to-person attack. It lets an intruder pose as a specialist attempting to resolve a problem with a hospital's medical system in order to obtain availability of the system and collect data. It enables them to potentially upload infection or find exploitable holes. In other circumstances, an attacker may pose as a visitor to a person, asking queries to discover a better understanding of the healthcare systems and devices in use.

TABLE 5.3

Various Attacks Employing Social Engineering with Solutions

Social Engineering Attack	Possible Cause(s)	Solutions
Social engineering	Poor training of employees	Training staff against baiting or pretexting
Reverse social engineering	There are no identification or verification procedures in place	Employees are being prepared to answer questions from strangers
Error debug	Different error questions give additional information	Limit appearing information

TABLE 5.4

Various Privacy Invasions with Solutions

Privacy Attack	Possible Reason(s)	Solutions
Traffic analysis	Unencrypted source and destination data A shortage of secure channels Poor encryption techniques	VPNs & proxies Non-linkability Pseudonyms
Identity/location Tracking	A shortage of secure channels Unencrypted location and identifying characteristics	Anonymity Non-linkability Pseudonyms

b. **Error Debugging Attacks:** These are frequently triggered by incorrect error handling, which makes medical systems open to a variety of security issues. Internal errors aimed at healthcare web servers, application servers, and web app settings can lead to information dumps, stacking traces, and error messages being shown to the attacker. A system call crash, a connection timeout, or unavailable databases are the most common outcomes. This uses a lot of resources and generates a lot of connection problems, which restricts and interrupts patients' access to healthcare.

5.3.2.3 Privacy Invasion

One of the most difficult challenges in IoMT is ensuring patient privacy. Patients' privacy is primarily concerned with preventing the revelation of their true identities, as well as their whereabouts and data. Patients must keep their personal information, such as their name, conduct, and history and current location, confidential [15,20]. In addition, they are identified and discussed in Table 5.4 in terms of privacy threats.

a. **Traffic Analysis Attacks:** TAAs are primarily concerned with patient privacy and data confidentiality. They are very risky attacks that involve intercepting and analyzing network activity to derive relevant information. This is because the actions of IoMT devices may provide enough information for an enemy to purposefully harm medical equipment. More specifically, traffic analysis may be used to target specific data that can be utilized to launch or support new social engineering assaults.

b. **Identity Theft Assault:** As part of his/her quest to steal the patient's identity, an attacker listens in on an IoMT device. In actuality, an intruder could be able to trace the activities of IoMT units. Tracing has the potential to reveal the patient's true identity as well as personal details. As a result, obtaining a patient's identify may jeopardize their privacy and perhaps their lives.

To protect a patient's privacy, the IP and MAC addresses must always be switched regularly to avoid identification exposure, spoofing, or DoS attacks. As a result, new techniques to address the enormous memory space challenge must be devised. An attacker listens in on an IoMT device in order to steal the patient's identification. In reality, an attacker may be able to follow IoMT device movements. This trail may expose the patient's genuine identity as well as personal information. As a result, acquiring a patient's identity may put their privacy, as well as their life, at jeopardy [13].

5.3.2.4 Message Validation and Data Security Threats

Integrity attacks target the validation of data or system by altering the messages that are being transferred. Injection attacks and data interception are two examples of attacks that can be used to attain this purpose. As a result, it is critical to safeguard and maintain data integrity to the greatest extent feasible [21].

a. **Message Alteration:** The intruder here is attempting to compromise the data reliability of the messages that have been transmitted. This occurs when an intruder manipulates received messages to achieve his or her objectives. As a result, doctors will make poor decisions that could endanger patients' health. One of the security solutions is to use a cryptographic function.

b. **Introducing Malicious Data:** An entity that is either lawful or appropriately permitted with the system initiates this form of assault. Sending a false message to the data center of the healthcare or clinicians can have dangerous consequences in the IoMT system and even result in fatal incidents. The purpose of this attack is to prevent authorized users from sending accurate and correct messages, instead infecting the network with fake messages. Messages should be verified to protect against such an attack.

c. **Malicious Script Injection:** These assaults build a fake update script framework that allows attackers to impersonate a reliable server for system backup. An attacker can get unauthorized access to any IoMT device and, in certain situations, construct a backdoor as a result of this.

d. **Spoofing Assault:** This might be used to mount a more advanced attack against medical devices or systems. Spoofing attacks make use of replicated data to gain unwanted access, whereas cloning attacks reproduce the spoofing data. The primary message integrity and authentication threats are summarized in Table 5.5.

5.3.2.4.1 Availability Attacks

Several assaults are conducted to weaken the functionality of healthcare systems in order to target their accessibility. As a result, availability attacks may target data or system accessibility.

TABLE 5.5

Solutions to Message Validation and Data Security Threats

Message Validation and Data Security Threats	Possible Cause(s)	Solutions
Message alteration Introducing malicious data Malicious script injection Spoofing	No protective method for data integrity or source authentication.	Keyed hash function (HMAC) Message authentication algorithms

5.3.2.4.2 Information Availability Threats

An intruder wants to compromise the accessibility of data from a message by discarding the exchanged message. When an attacker manipulates the incoming signal for his own gain, this occurs. This means hospital data centers and doctors miss important information about a patient's health [22].

5.3.2.4.3 System Availability Threats

The most common system availability assaults are mentioned here [22], and Table 5.6 summarizes them.

a. **Denial of Service (DoS) Attacks:** These are continually deployed to prevent real patients from receiving the correct medication to interrupt the accessibility of certain IoMT systems or devices, allowing nurses and physicians (GPs) to access medical information. They prevent recording. Due to service outages or interruptions, real-time data cannot be delivered or received [16].

b. **DDoS Assaults:** These assaults can be launched simultaneously from many locations throughout the world. This has a major influence on the accessibility of healthcare systems, as well as negatively impacting patients' lives owing to failure to reply in a timely manner.

c. **DE Authentication Assaults:** These assaults are often used to guarantee that only one de-authentication attack is carried out against a single medical device. It is sometimes used to initiate a bulk de-authentication procedure where all linked devices are turned off, either fully or partially. This technique can also collect handshakes, which can then be used to start cracking attacks against medical systems, devices, and even servers. This is prevented by frequency hopping and frequency shifting.

d. **Flooding Attacks:** These are predicated on overburdening and depleting the healthcare systems by introducing fake data and information into the system to flood it with fake information and data requests.

e. **Delay Assaults:** These cause rising communication flows to be delayed significantly. This gives you the option of re-transmitting them or not transmitting them at all once the time limit has passed.

TABLE 5.6

Various System Availability Threats with Solutions

Availability Attack	Possible Cause(s)	Solutions
Jamming	Access points and wireless IoMT devices are the targets	Beamforming, frequency hopping, and direct sequence spread spectrum
Denial of service	Inadequate backup devices	Backup devices
Distributed denial of service (DDoS)	Turning devices become bots by exploiting them	Detection of DDoS. To prevent becoming bots, increase the security of your devices
De-authentication	To launch a DoS or password cracking attack, the attacker captures a handshake	Firewalls, intrusion detection systems, encryption
Flood	False information injection overwhelms and depletes IoMT's resources	Timestamps, certificate authority, IDS
Delay	Overwhelms and prevents or significantly delays any medical information transmission	Firewalls, timestamps, IDS

TABLE 5.7

Solutions to Many Sorts of System Authentication Assaults

Authentication Attack	Possible Cause(s)	Solutions
Man-in-the-middle	Improper authentication technique	A multi-factor authentication mechanism
Masquerading	Improper authentication technique	A multi-factor authentication mechanism
Cracking	Improper authentication technique	A multi-factor authentication mechanism
Replay	The authentication protocol has a flaw	For each session connection, use a timestamp or a random number A multi-factor authentication mechanism
Dictionary	A single authentication factor and a weak password	A strong as well as lengthy password A secret key of adequate size
Brute force	A single authentication factor and a weak password	A strong as well as lengthy password A secret key of adequate size A multi-factor authentication mechanism
Rainbow table	Weak user names or password Short passwords	Long salt passwords
Birthday	Weak hashing	Secure hash algorithm
Session hijacking	Lack of encryption Unsecured channels	Encryption Sniffing filters

5.3.2.5 Tool/User Authentication Threats

Authentication assault attempts to obtain access to a system by bypassing passwords, which are considered the first and major line of security [23]. Attacks are usually successful in a variety of situations, and among other attack techniques listed in Table 5.7 can be used.

a. **MITM Attacks:** This is among the most popular identification attacks; it supervises and traces two lawful parties' interaction while modifying the data delivered. This attack might take the form of a passive or active assault. A passive assault occurs when the attacker simply retrieves and reads the communications sent and received between the two entities. On the other hand, if the attacker may alter the sent information without the awareness, it is termed an active attack.

b. **Brute Force Attacks:** These are usually built on an exhaustive investigation with all conceivable password permutations to crack a given medical password. The goal of such an assault is to steal patients' private medical information and credentials to commit fraud. Remote healthcare sensors and monitoring patients are among the most commonly targeted equipment.

c. **Masquerading Attack:** When a relay node on a wireless network is used for malicious reasons, it is called an attack. Such assaults can cause fake emergency medical warnings to be sent out on a frequent basis, impeding the availability of medical help. Furthermore, masquerading assaults might change a patient's health condition and result in the incorrect medicine being injected or overuse of pharmaceuticals, both of which can lead to death.

d. **Replay Attacks:** These alter the signal sent to other healthcare devices, particularly when an intruder acquires a high level of use of the system and the capacity to control its signals. By redirecting the transmitted data to a different site, the adversary can either steal or intercept it. Physical harm to a system, particularly medical systems, can be achieved in some instances. System communications are first captured and then later "replayed" to the receiving system. This results in unauthorized access and higher privileges on a medical system by stealing, leaking, or revealing critical information.

e. **Cracking Attacks:** To capture a handshake, cracking attacks use an assault on de-authentication. As a result, the intent access point is enticed to react with a handshake. After capturing the handshakes, a password recovery/cracking attack is launched for a specific healthcare system.

f. **Dictionary Attacks:** They are more common while attempting to obtain availability of a healthcare system. Assaults are more likely to succeed when safety precautions are fewer stringent than the safety precautions of a specified IoT system. To guess the password and get system access, such attacks use a vast number of dictionary terms. In reality, such an attack would need a significant amount of time and resources to carry out. Medical equipment with a weak security mechanism is frequently targeted by brute force assaults [24].

g. **Rainbow Table Attacks:** The password and its hash value using reverse engineering and a method known as "fault and try." It consists of a table of credentials and hash codes that are run until a match is made. Many solutions to this problem are provided in [24]. Salt passwords, on the other hand, can be an excellent way to protect against these types of assaults.

h. **Session Hijacking Assaults:** These assaults are carried out with the help of a session sniffer, which is a packet sniffer with the ability to modify, record, and read network traffic between parties involved. This applies to both individuals and gadgets. This exploit is capable of capturing a genuine session ID in reality.

i. **Birthday Attacks:** Users that rely on insecure hashing systems, where two distinct passwords might have the same hash, are likewise vulnerable to birthday attacks. Such a flaw might be simply exploited to get accessibility to any healthcare system without authorization. A hash function balancing was provided. SHA techniques, on the other hand, remain the strongest defense against such assaults.

5.3.2.6 *Malware Attacks*

Malware may target IoMT devices in a variety of ways, including viruses, spyware, backdoors, Trojans, and worms. This is due to a variety of factors, including their wireless and persistent Internet connection, as well as inadequate protection and monitoring. Malware is designed to take advantage of a software flaw, weakness, or a breach of security. As a consequence, it's feasible that a backdoor to a certain system or medical device may be created. It can potentially result in unauthorized admittance to the IoMT system and leak, change, or erasure of critical patient data. Attackers can utilize backdoors created by malware in IoMT devices to launch various forms of assaults or prohibit access to their services if the infection succeeds in creating them [14].

Malware protection is the most important enhanced safety for IoMT devices. This notion is demonstrated by cyber-attacks that leveraged IoT systems to create botnets. Another sort of virus assault that may seriously impair IoMT systems is malware, which prohibits them from completing their tasks. Advanced virus variants that use encryption or polymorphism approaches offer a severe hazard in this setting. As a result, antimalware software is essential to avoid malware assaults (Table 5.8).

TABLE 5.8

Various Malware Attacks and Their Remedies

Malware Attack	Possible Cause(s)	Solutions
Botnet	IoMT devices are a logical grouping of exploited Internet-connected gadgets	Pen testing, intrusion detection, and botnet detection solution (antimalware)
Worms & viruses	Relies on security flaws in computer networks	Intrusion detection, antivirus, pen testing
Spyware	Downloads via file-sharing sites or as part of other applications	Use antivirus software, update operating system, increase security and privacy, and intrusion detection
Remote access Trojan	Downloaded quietly through a program or software update	Updating antivirus software, blocking unused ports, and intrusion detection
Rootkit	Gains root access by exploiting and targeting whichever the kernel or even the user application space.	System setup that is appropriate, robust authentication, patching and configuration management, and anomaly detection
Ransomware	Paying ransoms, weak credentials, weak multi-factor	Updated antivirus, avoiding the use of private details, improving system security, and increasing awareness

5.3.3 IoMT Security Measures

It's a difficult endeavor to overcome the developing IoMT security issues and challenges. However, they may be mitigated by employing a variety of security measures, some of which are technological and others are non-technical.

5.3.3.1 Non-Technical Safety Precautions

This section focuses on the many non-technical safety precautions which can be used depending on the situation. This involves staff training and the protection of patients' confidential medical health records [25].

Raising awareness, performing technical training, and increasing the level of education might all be used to teach medical and IT personnel.

a. **Raising Awareness:** Raising awareness among healthcare staff, especially the IT staff, is critical and advised to recognize and distinguish an assault from internet-work operations. However, this is insufficient, because it is important to identify what defines a risk, weakness, or hazard [25]. This gives them the opportunity to spot a threat from afar. It also allows you to evaluate the likelihood and effect of a risk.

b. **Technical Training:** Raising awareness would not be enough; following the teaching phase, it's also critical to begin educating medical professionals and IT department employees.

c. **Improving Education Levels:** The present focus should be on raising education levels, particularly among individuals in the IT field. This is predicated on cyber-security and IT professionals being trained and educated on how to describe each attack and its target (confidentiality, availability, integrity, and/or authentication). There are two categories of attackers: insiders and outsiders. However, it's critical to consider the extent of an insider assault, as well as the probability of a distant or outsider strike. Then, teaching students how to estimate the probability of a hazard arising is another wonderful idea [21].

5.3.3.2 *Technological Security Procedures*

The technological security procedures should be implemented to provide an end-to-end protected IoMT system. As a result, the subsections that follow cover ways for guaranteeing IoMT system and data security.

Identification and Verification with Multiple Factors: It is critical to have a robust identity and verification process in place to prevent unauthorized entry to IoMT systems. Using biometric systems is the ultimate solution. A database is also required to properly and securely preserve biometric templates for future usage. However, attaining identification and verification necessitates the use of a number of biometric approaches, which may be classified as behavioral and physical biometric procedures [26].

Physical Biometric Techniques: Physically secure biometric strategies can be implemented and utilized to maintain and protect clinical privacy of patients while being resistant to internal threats.

Behavioral Biometric Technique: Hand geometry is a safe behavioral biometric approach that may be used for both verification and identification. In order to verify users, the data are compared to a database's collection of data stored. If a match is identified, accessibility will be allowed to the appropriate individuals. Accessibility will not be granted if this is not done. In fact, modern technology can tell the difference between a living and a dead hand. As a result, attackers are unable to fool the device and gain unwanted access.

Techniques for Multi-Factor Authentication: Authentication verifies the identity of the sender and receiver. In reality, authentication can be single-factor identification, which is insecure since it just uses a credential as a security mechanism. This may be two-step authentication, which requires additional security precautions in addition to the passcode to access a specific system [27]. As a result, authentication is critical in providing safety for the resources that are accessible on a specific network. Authentication could be centralized, in which nodes authenticated using a trusted third party, or dispersed, in which network entities authenticate one another with a predefined private key without depending on a trusted third party.

Furthermore, the energy source for a cryptographic key exchange authentication technique is external radio frequency instead of batteries [25]. This strategy may be used to keep unauthorized users out of your system. It is driven by the availability of extra channels, such as video and auditory channels, to produce a key which may be utilized to encrypt and protect body sensor transmissions in a network [26].

Availability Techniques: It is critical to sustain accessibility in the case of any potential disruption or stoppage of signals. Keeping the server's availability, on the other hand, necessitates the deployment of computational devices that serve as backup devices, and in the event of a system failure, backups and emergency response plans are verified.

Against Jamming: Jamming could take several forms, such as DoS, DDoS, and/or de-authentication. Many health services would be seriously impacted if jammer assaults occurred, especially if medical treatments were interrupted or disrupted. This disrupts and prevents interactions between healthcare equipment and doctors, resulting in missed revisions to patients' health information and, as a result, health concerns [28]. Further, if health services are affected due to a jamming attack, the initial responders will be late. This would raise the risk of a specific patient suffering from strokes, which might result in death. Various security measures must be built for this particular function in order to defeat any attempt that would attack the accessibility system. Maintaining backup computing medical equipment and servers, for example, is critical. In reality, medical gadgets have to be available all the time to ensure that all medical needs are met. Moreover, backup devices are able to react almost instantaneously and be triggered in the case of an emergency that affects the accessibility of a certain healthcare system. Extra security techniques, such as priority messages, channel surfing, and spatial retreat, should be considered and can be highly beneficial against wireless DoS assaults. This might be an important defense for healthcare devices, particularly in the IoMT space.

Honeypots: Honeypot systems are incredibly beneficial when it comes to identifying intruders, their aims, tools, and tactics. However, relying on static honeypot systems is difficult. As a result, the configuration of a dynamic honeypot system is necessary. Several honeypots are utilized in IoT and can be beneficial in the IoMT also. Using standard approaches, constructing honeypots for IoT nodes is difficult. As a result, an automated and intelligent method of collecting probable replies utilizing a scanner and leveraging machine methodology to understand the proper conduct during an encounter with an intruder.

Inexpensive IDS: IoMT devices are exposed to a variety of safety issues and assaults. The operations of IoMT devices should be measured and reviewed in order to secure IoMT systems from attackers. IDS are often the first line of defense in identifying assaults. The types of IDS that could be used in IoMT systems include host-based IDS (HIDS) and network-based IDS (NIDS). To detect potentially dangerous activity, HIDS is coupled to a specific IoMT device; NIDS examines the data transmission of numerous IoMT devices in the aim of identifying potentially suspicious attacks. IDS should be used to protect IoMT network systems so that anomalous activity may be detected as soon as possible, and necessary steps can be taken to avoid any occurrence [29]. When compared to unusual case detection techniques, cryptography and specification-based detection techniques have a lower overhead. However, standard anomaly-based IDS are inefficient in the IoMT instance because of limited computational capacity and a large number of coupled devices.

This raises serious security issues about present IoMT installations in general, as well as the necessity for a strong and inexpensive intrusion detection system [29]. Because enormous volumes of data must be analyzed, the research and industry groups continue to face obstacles in building reliable and effective IDS for IoT systems. Cooperatives hybrid IDS that are inexpensive and hybrid deployment and detection approaches are suitable options for making IoT networks immune to various assaults, particularly zero-day attacks.

5.4 Discussion

Failure to install encryption may result in data being intercepted, modified, and even erased beyond recovery. As a result, encryption mechanisms, particularly dynamic encryption, should be executed to secure sensitive information and preserve its confidentiality and privacy [13]. Additionally, given the majority of assaults are the result of phishing or social engineering, funding must be disregarded to raise awareness, train medical personnel, and enhance their technical understanding to detect any potential phishing or social engineering attack. Additionally, enhanced training should be provided to IT professionals in order to manage, protect, and safeguard the confidentiality of saved critical secret healthcare data.

It is significant to mention that there is a huge degree of mistrust among patients, who are expressing consequence over their privacy, particularly since the latest breaches revealed confidential medical information and patient's data [17]. As a result, establishing trust is critical and should be prioritized. Furthermore, inexpensive security procedures for encryption and authentication are necessary to enable the safe transfer of actual healthcare data, particularly for resource-restricted smart medical devices. This necessitates striking an optimal balance among IoMT effectiveness and privacy and security measures [30].

5.4.1 Inexpensive Cryptographic Algorithms

Security is built on cryptography techniques that ensure data integrity, availability, and confidentiality, as well as non-repudiation and origin verification. Implementing privacy and security safeguards, on the other hand, imposes a significant burden for particular kinds of IoMT devices. Many similar efforts were given to decrease latency and the resources required for these countermeasures. In certain cases, such as patient monitoring and transferring surveillance information, medical information should be transferred in actual, without any delay. Furthermore, traditional algorithms would rapidly deplete the battery life of IoMT end points [27]. To resolve this concern, the cryptographic schemes employ a flexible structure instead of a static one, where the cryptosystem primitives transition with each new input message, requiring only few rounds to obtain the required level of security, while a static structure might well need many rounds. As per [24], the approach achieves the targeted goals while also providing a degree of security needed for IoMT.

5.4.2 Inexpensive Authentication Protocols

IoMT authentication algorithms include various cryptographic algorithms and a hash function. The design of an effective cryptographic technique for IoMT may result in a reduction in the computation's required delay and resources [31].

5.4.3 Security Architecture with Layer

a. **Accuracy Layer:** The accuracy of healthcare activities is significantly reliant on establishing a three-way trust and understanding between medical staff, applications, and patients.

Sub-Layer of Trust: It demands the employment of the most precise medical apps, which should be exceedingly exact in coincident with no tolerance for faults. Furthermore, digital medical equipment must be validated by a licensed authority that might or might not be affiliated with a trusted third party [32].

b. **Prevention Layer** is necessary to avoid any inside organization assault, and to lower the chance of a remote cyber-attack exposing the patients' medical information. This necessitates the implementation of appropriate privacy, authentication, and confidentiality techniques.

Authentication Sub-Layer: This necessitates the establishment of an authenticating system, which depends on a vibrant and changing credential, as well as a biometric approach that is particular to each individual, making any effort to hack patients' data exceedingly hard [32]. This might be widely used in clinical staff in order to develop the proper authentication protocol by generating the minimum access per staff role. Furthermore, while employing medical apps, user/device identification must be created to guarantee physical safety and avoid physical tampering [21]. Finally, between the hospitals and the patient, origin identification as well as information integrity must be ensured by depending on a recognized authority.

Privacy Sub-Layer necessitates keeping patients' privacy a top concern. This necessitates granting patients identity and untraceability by enabling them to utilize a private connection while connecting to medical applications or websites. In addition to classic privacy-preserving data mining methods such as secret sharing, differential privacy [33], and homomorphic encryption, healthcare IT professionals depend on data mining methods that preserve privacy based on cloud computing.

Data Confidentiality Sub-Layer: To protect against inactive attacks, this layer must be properly maintained. This necessitates the use of inexpensive cryptographic techniques, including the use of quantum cryptography to safeguard valuable assets.

c. **Defensive Layer:** In order to maintain a safe e-health atmosphere, early identification measures must be taken before any remedial steps can be implemented [22].

Detection Sub-Layer: This sub-layer necessitates the establishment and deployment of the most powerful and ultramodern antimalware and antivirus tools, as well as AI-based strategies connected to dynamic and hybrid IDS, active honeypots, and SIEM. This will provide a high percentage of early and accurate identification.

Correction Sub-Layer: This should be preserved as a line of defense to thwart and counter safety threats. These features improved dynamic IDS, vibrant and advanced firewalls, and validated data backup with other systems accessible for required computing needs.

d. **Machine Learning Techniques:** Machine learning (ML) is a subfield of AI that learns from data and experiences without being explicitly programmed [28]. For persistent data analysis and production of useful information, ML has the potential to be quite useful in the IoMT, particularly at edge devices like fog/cloud computing. In a number of IoMT applications, machine learning has been applied. Conventional security solutions are incapable of protecting the system from zero-day assaults, since it is extremely expensive for IHS.

ML may be used in a variety of measures to fix the IoMT security breaches. Advanced machine learning algorithms can learn from enormous amounts of data collected by the IoMT, allowing for the discovery of new attack tendencies. As a result, data-hungry machine learning approaches may be fine-tuned using the IoMT. There have been three types of machine learning methods that may be utilized to solve IoMT security breaches. The supervised method is used for labeled data with predefined categories or labels. Classification and regression are the two types of supervised methods. These technologies, such as signature-based IDS, are used to identify attacks and malware. Unsupervised approaches are used to categorize the data based on their related properties because human labeling is not always possible. Semi-supervised method is just a newer ML field wherein specific data from a large set of learning data are labeled. Semi-supervised models could also be used to identify threats and prevent adversarial attacks on machine learning techniques [34].

In addition to the main classes of ML, a new branch of machine learning known as deep learning (DL), that is an upgraded form of a neural network, have recently emerged. In contrast to typical ML approaches that need the identification of extra features, DL may accomplish feature extraction/selection based on its learning process without the use of another approach. Unsupervised learning with auto-encoders is typically the best solution for anomaly detection. Because of its capacity to extract characteristics, the application of deep learning for safety solutions including such assault detection might be a long-lasting strategy for minor alterations.

The IoMT sensors, on the other hand, create a great deal of data at a rapid rate, resulting in big data [35]. Big data is defined as data that are either massive or complicated to be managed effectively by traditional technology and techniques. The three essential characteristics of big data, known as the 3Vs, are volume, variety, and velocity. It was discovered that big data and DL were employed together since DL performs better as the data increase. For security goals in industrialized IoT systems, big data and DL were recently combined with new blockchain and reinforcement learning techniques.

5.4.4 Detecting Sensor Anomalies in Medical Devices

The technique of detecting departures from the norm is known as anomaly detection. Anomalies such as this are linked to the occurrence of events that don't follow a predictable pattern. These anomalies are caused by unusual activity such as internal assault or injection of fake data, which causes the sensors to read incorrectly. The most often utilized technique for sensor security is the machine learning approach to anomaly detection. However, in addition to be approved within IoMT, this will need to be widely embraced.

Authorization and Authentication:
   Identity verification ensures that the IoMT user is legitimate, whereas authorization refers to the level of access granted to each user. They are the most useful methods for preventing cyber-attacks on the IoMT data's integrity and confidentiality. Cryptosystem, machine learning, and lightweight techniques may all be used in the identification and authorization process.

Intrusion and Malware Detection
   NIDS employing network traffic analysis is the most frequent type of intrusion detection. When the sensor and device log data are employed, that becomes HIDS. Both forms of malware detection employ measures to anomaly-based attacks and known attacks that differ from regular sensor data or network traffic. These IDS are identical in that they use an agent to gather information and a processing unit to identify and report intrusions. Signature-based techniques are good at detecting known attacks, but not so good at detecting new ones. Anomaly-based IDS [36], on the other hand, adapt from current data to discover unusual activity. As a result, they are capable of detecting new assaults, although they are less precise and computationally costly. IDS are most commonly used to counter modern threats such as DoS, DDoS, and other virus attacks.

5.5 Conclusions and Future Work

By connecting platforms and applications, IoT technologies have established a creative communication mechanism for those widespread healthcare systems. When IoT elements and infrastructures are connected with conventional medical systems, the concept of the IoMT emerges. However, because the IoMT works with a large variety of data assets that eventually influence a patient's health, there is a considerable security and privacy hazard within that IoT adoption. In this regard, IoT-based health systems must fulfil data security necessary conditions since they are transmitting highly sensitive patient data. Because end-to-end secrecy cannot be established without them, the security and privacy requirements of confidentiality, authenticity, accessibility, traceability, and non-repudiation must be satisfied in such systems. As a result, identifying possible risks, assaults, or issues associated with IoMT might help to increase awareness among all stakeholders involved in the IoT healthcare system.

Many studies were conducted to investigate those possible risks and assaults. As a result, in the IoT paradigm, we hope to show more extensive danger dispersion across various levels. Furthermore, the research findings show that the network layer has been the most sensitive layer to a variety of privacy and security risks and assaults, with the applications layer coming in second. Furthermore, it was discovered that DoS is the most prevalent risk in the network layer. We discovered that wearable technology seems to be the most acknowledged sensor device segment in the IOMT throughout our research. Furthermore, sensing devices have been highlighted as the most major element of IoT technology for tracking and monitoring bodily characteristics and motions. In summary, the purpose of this chapter is to optimize the linkages between various techniques as well as non-technical approaches in order to create a far more complex, safe, and effective system within all IoMT areas.

5.5.1 Future Research Directions

With the rise of other new advancements, including such cloud computing, there are several promising future research prospects that the IoMT privacy and security study community has yet to fully utilize. The following are some research directions that might be utilized to enhance IHS privacy and security.

a. **Blockchain:** Blockchain is a technology that is used to safely store financial blockchain data in a different location, where "blocks" are linked [37]. It would be focused on medical data distributed across medical servers, giving IHS enhanced privacy and security protection. However, in order to construct blocks, blockchain demands a substantial amount of processing resources on the platforms, which is not feasible on resource-constrained IoMT devices. Blockchain might be used to secure electronic medical record stored on medical servers. MedRec, for example, is a pioneering study on the use of blockchain for permission management and medical access to data.

b. **Artificial Intelligence:** In nearly every field, including network security, deep learning, and ML have become the most prominent study areas. In recent years, several ML-based intrusion detection approaches have been suggested, and they may be used in IHS as well. Because deep learning algorithms are widely being used in medical servers for disease diagnosing, the usage of such technologies for system privacy and security should also be considered. In one study, DL networks for intermediate assault detection were used to analyze PHI in various levels of IoMT systems.

c. **Security Assessment:** Adversarial assessment is one of the techniques researchers use to analyze the degree of security in their study. However, these adversarial analyses are not dependent on the same assumptions and concepts; therefore, they cannot be compared. As a result, the IoMT privacy and security community must build a methodology for measuring the privacy and security degree of security research.

References

1. Sengupta, J., Ruj, S., Bit, S.D., A comprehensive survey on attacks, security issues and blockchain solutions for IoT and IIoT, *Journal of Network and Computer Applications*, 149, 1–20, 2020.
2. Xiao, L., et al., PHY-layer spoofing detection with reinforcement learning in wireless networks, *IEEE Transactions on Vehicular Technology*, 65 (12), 10037–10047, 2016.
3. Kadhim, K. T. An overview of patient's health status monitoring system based on internet of things (IoT), *Wireless Personal Communications*, 114, 1–28, 2020.
4. Williams, P.A., McCauley, V., Always connected: The security challenges of the healthcare Internet of Things, *Proceedings of IEEE 3rd World Forum Internet Things (WF-IoT)*, 30–35, 2016.
5. Yeh, K.H., A secure IoT-based healthcare system with body sensor networks, *IEEE Access*, 4, 10288–10299, 2016.
6. EsatAnkaralı, Z., et al., Physical layer security for wireless implantable medical devices. *Computer Aided Modelling and Design of Communication Links and Networks (CAMAD), IEEE 20th International Workshop*, 144–147, 2015.
7. Chakravorty, R., A programmable service architecture for mobile medical care, *Proceedings 4th Annual IEEE International Conference Pervasive Computer Communication Workshops (PerCom Workshops)*, 1–5, 2006.
8. Sahi, M.A., et al., Privacy preservation in e-healthcare environments: State of the art and future directions, *IEEE Access*, 6, 464–478, 2017.
9. Moalla, R., et al., Risk analysis study of its communication architecture. *Network of the Future (NOF), Third International Conference*, 1–5, 2012.
10. Deogirikar, J., Vidhate A., Security attacks in IoT: A survey. *International Conference on I-SMAC (IoT in Social, Mobile, Analytics and Cloud) (I-SMAC)*, 32–37, 2017.
11. Mandy, D., et al., An overview of steganography techniques applied to the protection of biometric data. *Multimedia Tools and Applications*, 77(13), 17333–17373, 2018.
12. Karmakar, K.K., et al., Towards a security enhanced virtualised network infrastructure for internet of medical things (IoMT), *6th IEEE Conference on Network Softwarization (NetSoft)*, 257–261, 2020.
13. Somasundaram, R., Thirugnanam, M., Review of security challenges in healthcare internet of things, *Wireless Networks*, 1–7, 2020.
14. Tahsien, S. M., Karimipour, H., Spachos, P., Machine learning based solutions for security of Internet of Things (IoT): A survey, *Journal of Network and Computer Applications*, 161, 1–18, 2020.
15. Kumar, M., Chand, S., A secure and efficient cloud-centric internet of-medical-things-enabled smart healthcare system with public verifiability, *IEEE Internet of Things Journal*, 7(10), 10650–10659, 2020.
16. Elhoseny, M., et al., Secure medical data transmission model for IoT-based healthcare systems, *IEEE Access*, 6, 20596–20608, 2018.

17. Mahendran, R.K., Velusamy, P., A secure fuzzy extractor based biometric key authentication scheme for body sensor network in internet of medical things, *Computer Communications*, 153, 545–552, 2020.
18. Cheng, X., et al., Secure identity authentication of community medical internet of things, *IEEE Access*, 7, 115966–115977, 2019.
19. Yaacoub, J.P.A., et al., Securing internet of medical things systems: Limitations, issues and recommendations, *Future Generation Computer Systems*, 105, 581–606, 2020.
20. Vyas A., Pal, S., Preventing security and privacy attacks in WBAN, *Handbook of Computer Networks and Cyber Security*, Gupta, B., Perez, G., Agrawal, D., Gupta, D., Eds. Cham: Springer, 201–225, 2020.
21. Wang, J., et al., An efficient and privacy-preserving outsourced support vector machine training for internet of medical things, *IEEE Internet of Things Journal*, 8(1), 458–473, 2020.
22. Alsubaei, F., et al., IoMT-Saf: Internet of medical things security assessment framework, *Internet of Things*, 8, 100123–100156, 2019.
23. Zhang, J., Liu, H., Ni, L., A secure energy-saving communication and encrypted storage model based on RC4 for EHR, *IEEE Access*, 8, 38995–39012, 2020.
24. Alassaf, N., Gutub, A., Parah, S.A., Ghamdi, M. AL., Enhancing speed of simon: A light-weight-cryptographic algorithm for IoT applications, *Multimedia Tools and Applications*, 78(23), 32633–32657, 2019.
25. Xu, Z., et al., A lightweight mutual authentication and key agreement scheme for medical internet of things, *IEEE Access*, 7, 53922–53931, 2019.
26. Sun, J., et al., Lightweight and privacy-aware fine-grained access control for IoT oriented smart health, *IEEE Internet of Things Journal*, 7(7), 6566–6575, 2020.
27. Lu X., Cheng, X., A secure and lightweight data sharing scheme for internet of medical things, *IEEE Access*, 8, 5022–5030, 2019.
28. Ali, Z., M. S. Hossain, G. Muhammad, A. K. Sangaiah, An intelligent healthcare system for detection and classification to discriminate vocal fold disorders, *Future Generation Computer Systems*, 85, 19–28, 2018.
29. Abdaoui, A., Secure medical treatment with deep learning on embedded board. *Energy Efficiency of Medical Devices and Healthcare Applications*, 131–151, 2020.
30. Abhishek, N, V., Lim, T.J., Sikdar, B., Tandon, A., An intrusion detection system for detecting compromised gateways in clustered IoT networks. *IEEE International Workshop Technical Committee on Communications Quality and Reliability (CQR)*. Piscataway, NJ: IEEE, 1–6, 2018.
31. Aghili, S.F., Mala, H., Shojafar, M., Peris-Lopez, P., Lightweight three-factor authentication, access control and ownership transfer scheme for e-health systems in IoT. *Future Generation Computer Systems*, 96(1), 410–424, 2019.
32. Ahad, A., Tahir, M., Yau, K.A., 5G-based smart healthcare network: architecture, taxonomy, challenges and future research directions. *IEEE Access*, 7, 100747–100762, 2019.
33. Ahmad, B., Jian, W., Ali, Z.A., Tanvir, S., Khan, M.S.A., Hybrid anomaly detection by using clustering for wireless sensor network, *Wireless Personal Communications*, 106(4), 1841–1853, 2019.
34. Thuemmler, C., Bai, C., *Health 4.0: How Virtualization and Big Data Are Revolutionizing Healthcare*. Springer, 2017.
35. Huang, H., Gong, T., Ye, N., Wang, R., Dou, Y., Private and secured medical data transmission and analysis for wireless sensing healthcare system, *IEEE Transactions on Industrial Informatics*, 13(3), 1227–1237, 2017.
36. Nguyen, D.C., Pathirana, P.N., Ding, M., Seneviratne, A., Blockchain for secure EHRs sharing of mobile cloud-based e-health systems, *IEEE Access*, 7, 66792–66806, 2019.
37. Griggs, K.N., Healthcare blockchain system using smart contracts for secure automated remote patient monitoring, *Journal of Medical System*, 42, 130, 2018.

6

Cyber-Security Threats to IoMT-Enabled Healthcare Systems

S. Roobini
SNS College of Technology

M. Kavitha and M. Sujaritha
Sri Krishna College of Engineering and Technology

D. Rajesh Kumar
Galgotias University

CONTENTS

DOI: 10.1201/9781003256243-6

6.1 Introduction

In recent decades, the quality and scope of medical services provided by the traditional medical model have failed to meet the needs of patients. From a typical specialist or hospital-centered strategy to a patient-centered one, the smart healthcare system has experienced a fast shift. The Internet of Medical Things (IoMT) is an instrument with the intention of being allied together with a vital component of these technologies to provide a reliable outcome in the enlargement of the smart healthcare system concerning the efficiency of information processing appliances [1]. The IoMT is a network of hospital equipment along with apps to facilitate exchange in sequence and communicate with on-line computer networks to provide care.

The IoMT is realizing its full potential by utilizing smart objects that incorporate a variety of sensors and actuators in order to comprehend data via intrinsic networking capabilities and connect with all accessible options [2]. These gadgets can connect to open network services and interact with people. The network of gadgets plays a significant role in patients' and physicians' comfort and mental presence. It is a system that allows networked systems, applications, and devices to interact with one another, which will facilitate the monitoring and recording of the patient's health by doctors, which is very crucial medical information.

Recently, many technologies have been connected to current health services and medical resources to take advantage of the benefits of IoMT for the establishment of SHS. Because of the IoMT's incorporation with medical equipment, it is now feasible to transform patient healthcare from a reactive to a proactive system. Accessibility, cheaper costs, faster deployment, and greater efficiency are all advantages of connecting the IoMT with smart healthcare systems. Accessibility is improved by permitting physicians access to patients' actual data and allowing doctors and victims to access information more rapidly [3].

Since the IoMT is primarily used to capture very sensitive individual health data, the IoMT's security and privacy are important in preserving the patient's life, which may otherwise take a negative influence on the patient's health [4]. The following are the primary parameters of the smart healthcare monitoring platform. Data security is one of the most crucial aspects of privacy. Low latency [5] is an operational necessity for connectivity, and information authentication is a critical aspect with associated access control. Finally, the platform's accuracy is improved by the capacity to share data. At the very least, these critical criteria must be included in the smart healthcare tracking platform that will be developed. This vital demand has motivated advancements in IoMT technology that have consistently made remarkable strides in addressing security and privacy issues in IoMT.

6.1.1 Where Did IoMT Comes from

i. In every way, IoMT is a natural extension of the Internet of things. Healthcare-specific applications were created for such groundbreaking technology because it was unavoidable [6]. Remote telemetry may now be employed in a far larger spectrum of medical operations and practices due to downsizing, allowing us to be more precise, data-driven, and responsive in our biological systems. IoMT has sparked a surge in interest, and hospitals are trying to keep up [7]. To improve patient care, hospitals are keen to incorporate these gadgets into their operations. However, by doing so, they're opening up new attack vectors and putting hospitals in danger.

ii. Network devices tender a backdoor to the medical IT network, which is not effectively tenable and monitored, and may provide dreadful actors with a direct route to the healthcare industry's fragile. In reality, network-based vulnerabilities pose a significant risk of being exploited to catastrophic effect [8–11]. Big data insights, when combined with an elegant bioinformatics model and a large enough sample size, can lead to life-saving and improving medical treatment outcomes [12]. A continuous feedback loop allows healthcare practitioners to select the best treatment plan and alter it as needed. These compact and transportable gadgets broaden the number of disorders that can be treated remotely, making an emergency situation easier for patients and alleviating pressure on already-strapped hospitals and clinics.

iii. For an emergency, expert doctors may be reassigned to higher-priority treatments and only then return to the patient. Furthermore, electronic health records may be rapidly loaded with everything from a patient's medical history to diagnosis, prescriptions, and test results, obviating the need for yet another costly human intervention [9]. To extract data, modern technologies such as advanced analytics or artificial intelligence can be used on medical equipment and other technological assets and used to optimize operational and improve patient care.

6.1.2 Heart of the Problem

A healthcare IT system has the potential to open up an embarrassment of options in diagnosing, treating, and maintaining a patient's health and wellbeing, as well as hold the key to decreasing costs while increasing quality of care. However, as the number of connected devices rises, so does the risk of cyber-security. The importance of cyber-security is a security breach, a serious problem in healthcare and with an increase in high-profile hacks and hospital's rising reliance on IoMT equipment [10]. Attacks against healthcare organizations are common and sophisticated, posing a danger to patient confidentiality and safety. As a result of previous assaults, lawmakers have underlined the importance of cyber-security concerns impacting older software and infrastructures, notably ransomware. Manufacturers have no motivation to include medical device security requirements in their contract wording since there are no repercussions and no legal ramifications. The aftermath, which may involve severe financial penalties and a ruined image, is then left to healthcare organizations [13].

In the end, hackers are driven by a desire to profit from the theft of patient medical information. Treatment information can be used to impersonate someone, obtain free healthcare, or make false claims. A normal healthcare patient record contains certain information regarding their personal and medical history. A thief can create credit accounts or request for medical treatment with such a large amount of personal information [17]. While a person's financial identity may be restored completely, healthcare data breaches have a far more intimate and long-term impact on victims. Cybercriminals benefit from the healthcare business in a number of ways. But, in the end, it's their capacity to monetize that determines their success.

6.1.3 Motivations

A major issue in the healthcare industry is the constant exposure to cyber-security threats. These problems can stem from malware that mediates the system's integrity and disrupts patient's safety and treatment. IoMT devices will cause greater prob-lems due to technical shortcomings, and poor practices for patients and caregivers. Cybercriminals are constantly searching for new approaches to gain access to sensi-tive data. One of the main inspirations of cybercriminals is to make money, and the healthcare industry must act aggressively to mitigate this problem [17]. Cyber-security is probably more important for healthcare patients. Criminals can utilize the leaked information to be auctioned on the black market, which is then utilized for data crimes. Only by providing a full perspective of all assets in the medical ecosystem can total visibility and control be achieved, as it uses wired, wireless, and advanced technolo-gies [9]. As it works toward the final objective of treating additional patients with less health resources, it must consider security in order to effectively offer the required high-quality treatment.

6.1.4 Contributions

The study is unique in that it examines and analyzes all precautions and confidential-ity issues allied to medical IoMT devices in depth. The research also looks at existing lightweight security solutions, together with both computational and non-cryptographic approaches. Furthermore, a review teaches numerous lessons, and as a result, several rec-ommendations are made to make medical IoMT systems more secure and safe to deploy and utilize [12].

This chapter is organized as follows: Section 6.2 provides a framework of IoMT with its perspective and future, and it elaborates on communications, nature, and types of devices, and also application and service domains. In Section 6.3, the IoMT challenges, concerns, and risks are provided. Section 6.4 outlines cyber-attacks in IoMT and gives details on the nature of attackers and targeted security aspects of IoMT such as confidentiality, authen-tication, privacy, authorization, and availability. Section 6.5 describes the methodologies of the existing and proposed system with their comparative accuracies. Finally, Section 6.6concludes the chapter.

6.2 IoMT Framework, Perception, and Future

This segment presents the forms of IoMT devices together with the nature of medical devices. Then, protocols and application domains of IoMT are discussed, as well as advan-tages provided by IoMT devices. The framework is mentioned in Figure 6.1.

6.2.1 Devices of IoMT

1. Medical devices are classified based on their functions. In actuality, some of them are available as medical equipment, and hospitals employ them for clever remote monitoring in real time. Fitness trackers, blood pressure monitors, and glucose monitors are examples of medically advanced devices [12]. The aging population in affluent countries, a far more complex and appropriate healthcare system is required. IoMT is one of the generally essential solutions created to meet ever-increasing wants and demands.

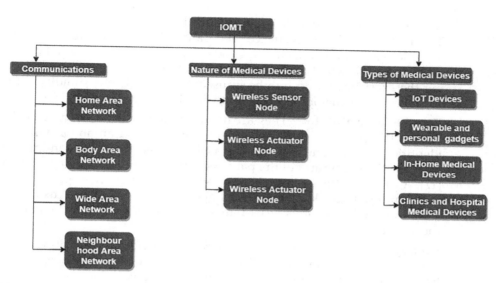

FIGURE 6.1
The framework of IoMT.

2. According to the World Health Organization, IoMT allows patients to move around more freely, resulting in a decrease in the number of patients in a hospital undergoing examinations. Hospitals may now remotely monitor diabetic patients [13]. These gadgets can be implanted, worn, or utilized in a variety of ways and can be carried in the hand. Furthermore, some devices are suitable for use at home, while others are intended for use in clinics.

a. **Devices for IoT:** There are numerous types of IoT devices available for use in medical applications. Devices on which an attack takes place are the doorbell, security camera, baby monitor, thermostat, and webcam.

   – **Wearable and Personal Gadgets:** Wearable and personal gadgets include medical gadgets that collect data and improve the health of patients within an instant at a lower rate. Wearable gadgets include heart rate monitors, healthcare monitoring wearables, portable diagnostic devices, ring-type monitoring devices, and nanosensors, which are examples of wearable devices [10]. Tele-home healthcare is getting increasingly common as the population ages and illnesses develop. Some of these smart devices are detailed in depth in the following sections.

   – **Suitability Devices:** They are hand-me-downs to help patients have a strong existence to maintain and enhance their health. A regular training plan that changes and is dependent on one's capacity and physical condition achieves this. Smart blood pressure devices are widely used [11]. They're used to monitor a patient's blood pressure in real time. These campaigns track changes in blood pressure caused by the norm in order to spot and report abnormalities in real time.

   – **Glucose Level Devices:** Patients with diabetes types I and II can employ sophisticated glucose level gadgets to monitor and measure their blood sugar levels. Maintaining the proper insulin level is critical for patient safety. This minimizes the consequences and dangers of unexpectedly higher or lower levels [19]. When insulin is produced, impulses are recorded on the insulin pump's actuators on the occasion of an insulin reduction, allowing the exact insulin dose to be infused. The spinal cord stimulator, for example, is an actuator that is placed to provide the patient with pain relief.

- **Emotion Rate Monitors:** They are hand-me-downs in a variety of medicinal settings which can save patients' lives. When an abnormality is identified, while some of the modern campaigns are able to continually examine the heart rates of the patients, others can just relay urgent data [17]. As a result, the major goal of these devices' is to arrive just in time before a heart attack occurs. BANs and wearable wireless sensor networks, as well as various heart rate monitoring devices, may be among these gadgets.

- **Dietary Devices:** They are being used to help individuals with eating problems maintain a balanced diet. They are mostly used by obese people who have difficulty sticking to those on a diet or who have lost track of their dietary limits. In reality, smart diet gadgets that exist mostly replace paper-based diets. Such gadgets would employ smart diet software to provide consumers with automatic updates about their daily meals, which would include various nutrition components.

b. **In-Home Medical Devices:** In hospitals, healthcare practitioners rely entirely on the necessary equipment to communicate [10]. Ventilators, infusion pumps, and dialysis machines are examples of in-home medical devices that are now used outside of a hospital or clinic. It has equipment such as first-aid, long-lasting medical, feeding, treatment, breathing, voiding, and inhalation, and also test kits.

c. **Medical Devices in Clinics and Hospital:** Hospitals have to be organized for any emergency or disaster in any respect of time, whether or not it's life-threatening [20]. As a result, for you to provide the finest remedy for patients, scientific devices and employees have to be well-organized. Donations of scientific devices are critical in this situation. Defibrillators, anesthetic equipment, patient monitoring, electrocardiogram (EKG) machines, surgery tables, and blank medical gadgets are just a few examples.

6.2.2 Application and Service Domains of IoMT

Despite the limits of the IoMT domain, it has a number of healthcare applications that benefit from it [1]. The expense of therapeutic concern is minimized through a decrease in the number of surgery appointments. Another benefit of IoMT is that it improves patients' health and quality of life. The capacity to access the imperative genuine analysis of a diagnosis for early detection, medication prescriptions, and injections via a wearable device is all possible. Doctors, nurses, medical supplies, and receptionists will be part of the IoMT future, which will benefit from devices and software. The general public, on the other hand, is concerned about the IoMT systems' needed protection, confidentiality, hope, and precision.

a. **Smart Medical Technology:** This term refers to how paramedics are currently deploying and using smart medical gadgets and supplies to provide rapid medical treatment to those in need. Drones are being used to do such a task. To increase accuracy, medical technologies based on AI are also being used [14]. In the beginning, medical drones were employed to respond to cardiac arrest cases by applying a predictive risk assessment technique to remotely convey crucial signals to the hospital [21]. This supports the use of smart medical robots in hospital settings to perform surgical procedures. A medicinal technology based on fundamental/amplified realism (VR/AR) with artificial intelligence (AI) has been used in a number of medical applications. Virtual education, remedial mentoring, and cardiopulmonary learning in cardiac surgery (CPR) are just a few of the realistic procedures that virtual reality may be used for. To increase accuracy, medical technologies based on AI are also being used.

b. **Smart Receptionist:** The drones would be commanded to fly to certain areas, saving time and, as a result, lives. Another IoMT innovation is the smart receptionist, which is a medical robot that can, before referring a patient to the appropriate medical department, think about and grasp a specific medical or urgent concern [17]. Those machines would also be able to take calls and schedule appointments for patients, categorizing them as imperative or non-urgent. Statistical or machine learning approaches are used to categorize data that might be used.

c. **Personal Emergency Response Systems:** They are a type of personal emergency response system. Patients and clinicians are increasingly being alerted by remotely conveying critical transmissions to the hospital using a risk assessment that predicts the future in real time of any unexpected medical incident [18]. To increase accuracy and reaction speed, PERS is currently undergoing a transformation to become location-based. In particular, the smart belt that is both active and protective can be worn around the waist of a patient and transmits real-time data through Bluetooth and AI.

d. **Ingestible Cameras:** Ingestible cameras are tiny, minimal capsules to facilitate a patient's being capable of getting through instantaneous imaging vision surveillance of internal organs for the early identification of chronic illnesses and cancer. A data recorder capsule, an endoscopic optical scanning device, and a hydrogel device that can be ingested are among the ingestible products on display [18]. For assessment, ingestible devices use the diagnostics toolbox and a tracking or recording system.

e. **Intelligent Experts:** The long-term objective will be developing an efficient smart robotic system as an idea that is capable of performing an activity as well as the duties of a genuine professional. Certain individuals also stated their concerns, whereas some have stated that chatting with this is more comfortable in talking toward a medical automaton regarding personal therapeutic problems, it is to talk to a real doctor [19].

f. **Smart Nurses:** In addition to primary medical activities, secondary medical functions, such as functioning as a nurse, will be possible for smart medical robots. In many situations, they may act as a nurse smart assistant to make the nurse's job simpler [19]. According to medical circumstances and needs, robots will be used to undertake a secondary and/or supportive medical duty.

6.3 IoMT Challenges, Risks, and Concerns

In this technology's evolution, there are a number of obstacles that might hamper innovative healthcare applications [16]. Security and privacy are the most pressing concerns. The primary security difficulties, obstacles, and hazards involved with the installation of IoMT systems are discussed in the following section.

6.3.1 IoMT Challenges

The issues occurred when medical equipment was connected to IoT systems as soon as they were available. The lack of consistency is a major problem. Medical equipment must be standardized in order to interact with one another, and manufacturers must take appropriate security precautions to avoid hacking. Higher protection, efficiency, scalability, consistency, and effectiveness would result as a result of this [15]. Many of these problems are, in fact, principally related to, but not limited to, various IoMT security flaws, for example memory constraints, computational and energy constraints, and tamper resistance. Embedded software limitation and dynamic security patch are two types of software. Then there's the network, which has to do with device diversity, communication medium, multi-protocol networking, scalability, and dynamic network topology [14].

Furthermore, vital interoperability between wide ranges of device types is frequently achieved by clever solutions, sometimes by staff with just rudimentary cyber-security education and expertise, putting the company at danger. There are certain challenges to be considered: Cyber-security ownership difficulties arise in a multi-player system.

- It's a game with the devil to service, patch, and otherwise update key equipment.
- Medical equipment is frequently linked to legacy infrastructure that has evolved over time.
- Despite manufacturers' best efforts, some linked medical devices are already vulnerable when they are introduced.
- Clinical personnel are frequently unaware of fundamental cyber-security issues.

6.3.2 Risks within IoMT

The incorporation of IoMT systems into the healthcare industry sector has a variety of hazards, which are outlined below [16–18]:

- Personal information sharing is required by the hospital's norm, which might have a substantial influence on patients' medical situations.
- Unsatisfied or rogue medical professionals, in exchange for a bribe or as part of an organized crime operation, leak medical facts and details about the facility or the patients, jeopardizing patients' privacy.
- Nurses and physicians who lack training might endanger patients' lives, resulting in lasting disabilities or death.
- Accuracy is still a contentious topic, and it is still to blame for errors in medical procedures performed by specialized robots. This can have a significant impact on patients' life, resulting in impairments or death.

A novel approach to risk assessment will be desired toward determining the safety measures hazards of IoMT attack may be a tough mission.

6.3.3 IoMT Concerns

Concerns with IoMT, one of the three categories is stated with the broad community and is coupled toward issues of isolation, confidence, and protection [19].

a. **Concern for Privacy:** Even more so in terms of passive attack, traffic analysis creates privacy problems since it is possible to collect and expose in sequence in relation to individuals identity, as well as sensitive and personal data [13]. Because an adversary is capable of determining a patient's therapeutic data and medicinal concerns, this is a big hazard to patients, which can have dire consequences for patients' lives. Identity theft is another motive for invading patients' privacy by assaulting hospitals. The majority of these real-world assaults resulted in a violation of an individual's confidentiality throughout outflow or else exposure of personal data [17]. To recap, isolation entails supplementation rather than simply concealing sensitive and secret medical data. It also demands ambiguity, non-linkability, and non-observability.

- **Ambiguity:** When communicating, identity must not stand revealed and his identity should be concealed. Passive attacks, on the other hand, may observe what you do, but not who you are.

- **Non-Linkability:** Subjects, communications, events, and activities are examples of items of interest (IoI). Likelihood in such things was hidden according to the victim's point of view before and after observation should be the same.

- **Non-Observability:** The state of non-observability occurs when items of interest (IoI) are interchangeable since other IoI of the identical category. As a result, communications are indistinguishable from random noise. This would be whether a message has been transmitted between a sender and a recipient in any connection should be undetectable.

b. **Concern for Trust:** Patients' confidence is jeopardized when their privacy is violated [20]. Patients are growing suspicious of the concept of robots taking over human functions (doctor's office, nurse, and assistants). Inhabitants are more concerned with the presence of a therapeutic droid, technology, and perhaps a medicinal gadget monitor and regulate their health issues.

c. **Concern for Security:** IoMT devices are vulnerable to a number of wireless/network attacks since they rely on open wireless connections. Due to flaws in security mechanisms, an invader preserves to listen and record external and departing facts in real time. Most of the authentication techniques in IoMT devices have a loophole by which a competent attacker finds a way [21]. Another security risk is the possibility of gaining unauthorized access while going unnoticed because of the inability to detect and stop such attempts.

Medical devices can be hacked by botnets or zombies, resulting in physical assaults on victims. For example, an assault may conceivably alter a pharmaceutical dose that would kill or badly damage a patient. Furthermore, if seized by terrorists, IoMT devices might be utilized for targeted killing [22]. Furthermore, there is a detrimental impact on patients' psychological states since they may be afraid of them, resulting in a heart attack as a result of being surrounded by IoMT equipment. In order to secure and maintain security [24], as well as medics, medical equipment manufacturers must prioritize security. It provides defense against both passive and violent attacks that are essential for mitigating the primary IoMT security risks. As a result, the necessity for appropriate security measures and technologies is critical.

6.4 Cyber-Attacks Aligned with IoMT

Attack can be embattled, organized, or even coordinated to successfully carry out a cyber-attack, depending upon the assailant's talents, expertise, and skills [23–27]. In reality, the sort of malware employed to carry out the assault has an impact.

6.4.1 Features of Cyber-Attacks

It is critical to comprehend the features of an assault before detecting and classifying it. Each assault may be classed into one of five fundamental categories based on the kind, target, scope, capability, and effect of the action, as well as the purpose, objectives, and targets. More specifically, it is determined by the attacker's abilities, data, understanding, and accessible tackle and possessions [25].

a. **Nature of Attackers:** Internal, external, aggressive, passive, malicious, rational, well-organized, and well-coordinated are the characteristics of attackers. Different sorts of attackers may work together in some situations in order to prevent a more sophisticated cyber-attack within and outside the firm.

- **Internal versus External Attackers:** An internal attacker is frequently a coworker, such as a nurse or doctor, who is concerned about their patients' safety and privacy. In other cases, a secret agent disguised either like a physician or a specialist may have been able to circumvent altogether of an infirmary's safety measures in instruction to remove an enduring aimed at supporting or additional immoral purposes. External attackers may find it simpler to carry out cyber-attacks if inside attackers make it easy for them.

1. Malicious hackers are the most common external attackers, with the goal of getting enhanced unlawful exclusive access to the hospital's computer system. Worms, rootkits, and remote access Trojan assaults are the most common ways to do this [26]. The assault is frequently based on spear-phishing tactics, such as curriculum vitae, transmitting a corrupted long documents and or a kind of attachment. Once the computer has been a trapdoor or keystroke trackers are deployed and downloaded.

2. **Aggressive versus Passive Attackers:** A passive assailant seeks in the direction of stay undetected with being invisible in the background and without doing anything. Data collection method that might be exploited in the future, perhaps leading to an even more sophisticated cyber-attack [27]. Information is obtained by passive intruders while elements of a research procedure may collaborate with external or even inside attackers.

3. An active attacker, unlike a passive attacker, focuses on intercepting a source's and a target's communication. Without knowing the source and destination, such interceptions are carried out aggressively by changing, modifying, and deleting the conveyed data and information. When used to administer a greater dosage of a medicine to a patient, such an attack is extremely dangerous, endangering the patient's life.

- **Malicious versus Rational Attackers:** Malicious attackers have no precise aim in mind and aren't looking for certain outcomes. They initiate assaults solely because they are capable of doing so, with the goal of disrupting the Internet of things system. This can be accomplished by sending fake information to a data center in a certain geographic area. For example, attackers who were rational, on the other hand, have a definite target in mind that may be quite hazardous. To put it another way, they will be capricious and the passive class is frequently followed.

- **Well-Organized versus Well-Coordinated:** IoMT virtual assaults preserve as well-organized or well-coordinated. Coordination and coordination between insiders and outsiders are the foundations of coordinated assaults. Outsiders with a vested high level of remote admission to a certain therapeutic system access or privileges can use malware kinds to launch a coordinated attack. The attack strength is conducted to disturb medicinal processes by denying access to medical records, scheduling appointments, and then disrupting medical procedures done by the authorized therapeutic workers and patients.

b. **Target:** A targeted attack is usually carried out with the intention of assassination or terrorism. A patient or a facility may be targeted for a variety of reasons, including biased, intellectual, cultural, or virtuous motivations. Targeted device specifies IoMT systems, medical devices, medical servers, medical data, and patient's information [27].

c. **Impact:** The quantity of damage an assault produces, as well as its type and extent, is used to determine its impact. The impact can be low, moderate, and serious.

d. **Capacity:** The protection necessary to avoid, alleviate, or lessen the damage associated with capacity is referred to as capacity: detection, correction, non-correction, and non-detection.

e. **Scope:** The scope of an assault is determined by the size of the target region, which can be classified as small or big. The scale of a scope can be small, moderate, and large. Attackers frequently aim to spread their bad acts across a vast region. It is an excellent idea to raise the frequency of offenders, for instance hospitalized personnel.

6.4.2 IoMT Targeted Security Aspects

Various sorts of cyber-attacks appear to be jeopardizing IoMT security, which are categorized and characterized according to the security component that they target. The goal of this part is to look at the security assaults that focus on data security in IoMT, such as accessibility, privacy, and reliability. Figure 6.2 describes the types of attacks.

a. **Privacy Attack:** One of the most difficult difficulties in IoMT is ensuring patient privacy. Patients' privacy is primarily concerned with preventing the revelation of their true identities, as well as their whereabouts and data. This necessitates patients protecting their personal data, such as their identification, behavior, and history, as well as their current location.

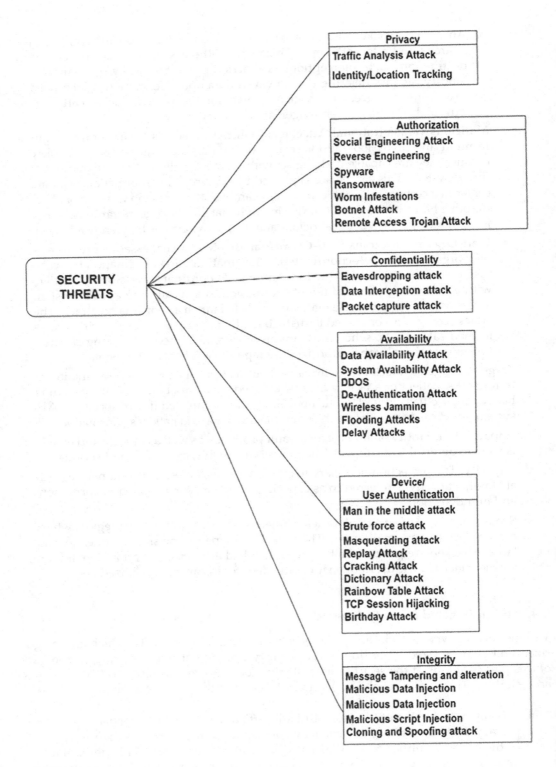

FIGURE 6.2
Security threats to IoMT.

- **Traffic Analysis Attack:** TAA primarily concerns the privacy of patients. This is because the operations of IoMT devices might possibly give enough information to allow an attacker to hurt medical equipment maliciously. More specifically, traffic analysis may be used to target particular information that could be utilized to help or instigate future social network attacks.

- **Attack on Identity or Location Tracking:** In an IoMT device, the hacker spies on it to determine the patient's identity (by linking the patient to a location). In reality, an attacker may be able to track the movements of IoMT devices. In addition to the patient's personal information, this trail can expose the patient's genuine identity. As a result, obtaining a patient's identity may jeopardize their privacy and perhaps their lives. To safeguard a patient's privacy, the device and network address must be altered on a regular basis to prevent identity theft and a denial of service (spoofing) attack.

  1. As a result, new methods must be developed to meet the enormous memory space problem. As a result, a certificate authority should provide each patient with a set of certified pseudonyms. The most well-known attack is the Sybil attack. The pseudonym collection may be used toward bogus communications to a data center while posing as various patients. This can include things such as fake traffic bottlenecks or false notifications which force a fictitious calamity. The primary objective of the authorities is to make certain that people's credentials and confidential material are preserved and authenticated throughout each interaction effort. System operators must intervene in the case of a problem. Yet, this entails knowing the user's identity (digital forensics).

b. **Availability Attack:** Different assaults are carried out with the intention of degrading the functioning of medical systems. As a result, data or system availability might be the target of availability attacks.

  - **Attack on Data Availability:** When an attacker tampers with incoming signals, it's known as signal tampering personal advantage, leading the hospital data center or physicians to miss vital information regarding a patient's health.

  - **System Availability Attack:** The most common type of assault is a denial of service (DoS) attack prevalent system availability assaults, which aim to prevent genuine patients from obtaining adequate drugs and disrupting the presence of a therapeutic IoMT device or gadget prevents assistants and physicians in obtaining health data. Due to the loss and suspension of service, real-time data cannot be delivered or received.

    - **Distributed Denial of Service Attack:** Such assaults are deadly assaults, launched concurrently from multiple geographical regions and nations. Due to the inability to respond promptly, this could consume an important influence on the availability of medical equipment and schemes, putting patients' lives in jeopardy.

    - **De-Authentication Assault:** These are assaults that are directed at single medical equipment. It may also be used to perform a bulk de-authentication operation that temporarily or permanently disables all linked devices. This method might be used to get illegal access to a medical system as well to record a handshake, which can subsequently be utilized in a cracking attempt.

- **Wireless Jamming:** The spate of cognitive dissonance attacks strikes that never stop are largely aimed against wireless networks, causing any connectivity to be disrupted. The nature of jamming assault varies on whether that is selective or non-selective and whether there are secure and non-secure channels.
- **Flooding Attack:** These attacks are centered on overpowering and straining the by utilizing the medical system's resources introducing fake data and information into the system, flooding it handling erroneous research and data inquiries.
  - **SYN Flooding Assault:** It is also known as half-open attacks, largely targeted against them. The aim is to bring down a clinical service by eliminating the cache backup host, permitting future threats to take the advantage of unprotected connectivity.
  - **Black Nurse Assault:** They are low-bandwidth ICMP attacks that use DoS attacks; firewalls operated by an effective central processing unit (CPU) burden should be targeted. LAN users, such as patients and medical workers, are unable to send Internet network traffic as a result of this assault.
  - **Delay Attack:** They cause significant delays in the delivery of high-priority messages. This gives you the option of re-transmitting them or not transmitting them at all once the time limit has passed.
c. **Device or User Authentication Attack:** Authentication attacks seek to expand access to a scheme by passing the first and most important barrier of defense passwords. For example, when a security code is too frail or too simple, or when it is stable. Brute force, dictionary, birthday, and rainbow table attacks are all possible encryption breaking.
  - **MIM Attack:** This is the most recurrent validation that affects data sent, oversees and analyzes the conversation between multiple competent individuals, and modifies data delivered. It might be a passive or active assault. A passive assault happens once the intruder just retrieves and records the conversations between the two entities. An active attack, on the other hand, arises whenever the intruder can edit, modify, and/or alter the presented facts and statistics without the app's notice.
  - **Brute Force Attack:** These threats include a thorough search for all possible password combinations in order to crack a given medical password. Remote medical sensors and patient monitoring are among the most commonly targeted equipment.
  - **Masquerading Attack:** It occurs while using a wireless network relay node is used for malicious purposes by a single attacker. Such an assault might repeatedly broadcast false warnings indicating an emergency medical condition, causing medical care to be unavailable. Furthermore, spoofing attacks may alter an individual's medical circumstance, resulting in injection of incorrect medication or the overuse of drugs, both of which might result in human life loss.
  - **Replay Attack:** Attacks that modify the control signal given to medical equipment are known as replay attacks, especially if an attacker has gained a high level of access and control over the system's signals. By diverting the sent information to a different site, the adversary can either steal or intercept it.

- **Cracking Attack:** It relies on a de-authentication attack to capture a handshake. As a result, the desired access point is enticed to react with a grasp. After capturing the connection, a keylogger assault is carried out by a real health platform or equipment. This enables the interchange of statistics and facts leakage exposure.

- **Dictionary Attack:** These threats are more common while attempting to obtain access to a healthcare system. Medical equipment with a weak security system is frequently the subject of brute force attacks.

- **Rainbow Table Attack:** It relies on a method known as "fault and try" and reverses engineering to attack the password as well as its hash value. It usually consists of a table containing passwords and hashes works through until a match is found.

- **TCP Session Hijacking:** It is another name for session takeover bouts, packet sniffer that can change, record, and read set of connections interchange between multiple individuals. This is true for both people and gadgets. In reality, this exploit is capable of capturing a legitimate session ID (SID).

- **Birthday Attack:** Users that rely on poor hashing systems, where two distinct passwords might have the same hash, are likewise vulnerable to birthday attacks. These weaknesses may simply exist oppressed to get access to any medical system that is not authorized. Balancing was given in a suggested hash function. Gathering information is required to ensure the secrecy of IoMT data and certain approaches continue to provide the best defense against such attacks.

d. **Data Confidentiality Attack:** Gathering information is needed to maintain the confidentiality of IoMT facts. Because of its open and public nature wireless connections, patients are increasingly exposed toward interception by privacy assaults (sniffing). As a result, personal and intimate data may be undermined, compromised, manipulated, or even theft. Data confidentiality includes eavesdrop, traffic analysis, as well as brute force assaults.

- **Eavesdropping Attack:** These attacks are classified based on how they gather facts; these assaults are subdivided as follows. The first is proactive eavesdropping, which includes scanning wireless access points for medical equipment that are related toward each one. The subsequent category is active dropping, in which an attacker could keep track of data transfer while it is being transmitted, by allowing them to capture more data more quickly and easily.

- **Data Interception Attack:** This attack occurs when MIM is used and promotes the attacker to seize fact and later retransmit. This enables the attacker to intercept and retransmit the Address Resolution Protocol (ARP) request until a handshake is captured. The encryption keys obtained through this handshake are then used in gaining unrestricted utilization of healthcare records and systems.

- **Packet Capture Attack:** It is also known as packet sniffing attacks and entails capturing and disclosing the contents of unencrypted medical data packets. A network monitoring software program such as Wireshark is a good example.

e. **Integrity and Authentication Attack:** Authenticity assaults focus here on capacity to change information delivered in order to jeopardize a platform for data's reliability. Injection threats and data interception are two examples of attacks that can be used to attain this purpose. As a result, it is critical to safeguard and maintain data integrity to the greatest extent feasible.

- **Attack against Message Tampering and Alteration:** The assailant's purpose is to weaken the information uprightness of communication existence swapped. Using a cryptographic keyed hash function as a message authentication technique is one of these security measures (HMAC).

- **Malicious Script Injection Attack:** These threats simulate using a legitimate server for system backup fraudulent update script system. This allows a single attacker to get illegal access to any IoMT device, as well as potentially introducing a trapdoor.

- **Cloning and Spoofing Attack:** Threats may be combined to create a more potent attack. Spoofing attacks use data that have been duplicated to gain unwanted access, whereas cloning attacks copy the faked data.

f. **Authorization Attack:** Adversaries may attack IoMT's weak authorization procedures to gain permissions without the need for a password. Authoritarianism might be used on IoMT devices as a result of users' lack of security training and awareness. As a result, a malicious actor may deceive and pose when trying to have accessibility to the person's health records equipment. Malware assaults are potentially possibly linked to IoMT devices by taking use of their inherent vulnerabilities, such as holes in endorsement processes. Authorization incorporates social engineering attacks, reverse engineering attacks, and malware attacks.

- **Social Engineering Attack:** People are manipulated using social engineering techniques such as baiting or pre-texting to get them to divulge information. In sort to hold out a computer security, this comprises passwords, names, IDs, and sensitive information. It appears that luring individuals may be accomplished more readily by depending on human emotions rather than exploiting a system's weakness. As a result, the attacker preys on people's passion or curiosity and distributes infected pornographic photographs to get access to medical systems or data.

- **Attack Using Reverse Engineering:** The assailant could posture as a specialist trying toward solve a problematic with a welfare scheme of an institution to gain information which has access to the system on a physical level. It enables to presumably deploy malicious software to find exploitable holes. In other circumstances, an attacker may pose as a visitor to a patient inquiring about medical systems and gadgets to have a better understanding in use.

- **Malware Attack:** The most common reason is incorrect error handling, which makes medical systems open to a variety of security issues. This requires a lot of resources and creates a lot of network overhead, which prevents and disrupts patients' access to medical treatments. Malware attacks come in a variety of forms:

  - **Spyware Attack:** Spyware's primary goal remains to get and collect data on patients in order to sell them on the dark web or transfer them to a third party.

  - **Attack by Ransomware:** In greatest cases of classic ransomware, the attackers exploit the user interface to remind the victim that they need to pay the ransom. A considerable majority of IoMT devices, however, lack a display interface. In this situation, the attackers try to find out their owners' e-mail addresses or hack the app that manages the hacked IoT devices. Because it is timely, vital, and reversible, IoT ransomware is effective. As a result, attackers select scenarios in which victims do not have the time or resources to restore the device's functionality or lessen the consequences of the ransomware. Ransomware has discovered that IoMT devices are attractive targets.

- **Worm Infestation:** They may self-propagate without the need for human involvement. They may have an influence on all data and device security services, resulting in significant data loss or life threats [28]. The dubbed Internet worm was a recent harmful Internet infection that targeted IoT devices. It is installed and deployed in opposition to IoMT devices to collect data, cause harm, or even destroy them. In the case of IoMT, unprotected devices might infest the entire medical system's security. Worms can be paired with a variety of different organisms of malware to propagate over the IoMT network, such as ransomware and botnets [32].

- **Attack by Botnet:** They might be used in DoS [29] or DDoS attacks to knock down the whole healthcare system. For example, the Mirai attack used malware to infect IoT devices in order to construct botnets and launch DDoS assaults against network servers, infrastructure, and other targets. As a result, these devices may be utilized to use bots to attack medical systems. It's worth mentioning that the Mirai [31] onslaught has evolved into new mutant forms and more powerful ones are being developed all the time.

- **Remote Access Trojan Attack:** RAT attacks are carried out by exploiting a medical system's vulnerability, flaw, or security lapse. It is mostly accomplished by circumventing the authentication procedure. The operation Shady RAT was the most well-known attack [30].

6.5 Methodologies

In the previous section, several attacks were discussed. Among them, botnet attack is analyzed by using deep learning models. A framework is developed for detecting botnet attacks by means of the N-BaIoT dataset.

6.5.1 Dataset Description

The N-BaIoT dataset was created by analyzing the network traffic of nine IoT devices in real time. There is both benign and malicious traffic in the data. The information is alienated into files for each device, with each file including a different sort of traffic, such as regular traffic or assaults. Using two relations of botnet occurrence programs from GitHub, ten different types of assaults were created (Mirai and Bashlite). The N-BaIoT dataset contains 115 characteristics, which are statistical analyses taken from packet traffic during distinct time eras.

The dataset covers the succeeding nine instruments with attack and standard traffic: doorbell (Danmini), doorbell (Ennio), baby monitor (Philips), security cam (Provision_PT-737E), security cam (XCS7–1002-WHT), web camera (Samsung), thermostat (Ecobee), security cam (Provision_PT-838), and security cam (XCS7–1003-WHT).

TABLE 6.1

Feature Information

Value	Statistics	Total Features
Size of packet	Mean and variance	Eight
No. of packets	Number	Four
Jitter packet	Mean, variance, and number	Four

TABLE 6.2

Entire Number of Attacks in Respective Device

Instrument	Benign	Mirai	Bashlite
Doorbell (Danmini)	49548	652100	316650
Security cam (XCS7–1003-WHT)	19528	514860	316438
Security cam (XCS7–1002-WHT)	46585	513248	303223
Web camera (Samsung)	52150	-	323072
Doorbell (Ennio)	39100	-	316400
Security cam (Provision_PT-838)	98514	429337	309040
Security cam (Provision_PT-737E)	62154	436010	330096
Thermostat (Ecobee)	13113	512133	310630
Baby monitor (Philips)	175240	610714	312723

6.5.1.1 Characteristic Details

The 115 structures result from a traditional 23 structures. The same set of 23 features extracted from five time gaps of the greatest fresh: 100 ms, 500 ms, 1.5 seconds, 10 seconds, and 1 minute. These structures can be calculated actual wild and incrementally and thus enable actual period discovery of mean sachets.

The foundation IP is cast off to path the entire mass. The foundation MAC-IP feature allows you to identify circulation coming from various entries and traffic coming from faked IP addresses. The basis and end point TCP or UDP have n-facts that determine the sockets. Table 6.2 demonstrates the entire number of attacks in respective device.

6.5.2 Existing Method

In the existing method, botnet identification is currently carried out by studying performance disparities between different types of IoMT devices and deep learning models. Using the dataset divided function of Scikit-learn, randomly split the exercise and difficult models by 71–31 among the N-BaIoT dataset samples; the exercise and difficult circles are self-governing of, respectively, the other. Practice 25% of the exercise traditionally as authentication set to avoid over-fitting. We employ the most commonly available deep learning methods as botnet training models. CNN and RNN [6,9] with long short-term memory (LSTM) [3] were used as DL copies. Multiclass organization as well as two arrangements in this part. The deep learning models are used to train the model, which is dependent on the dataset acquired from each device. Then, during training, compute the validation loss to see if the validation loss does not rise as the training loss reduces. Then, using the F1-score measures, assess the model's performance.

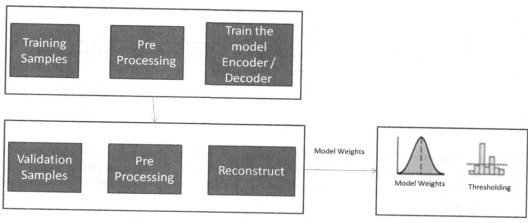

FIGURE 6.3
Autoencoder model.

6.5.3 Proposed Method

An efficient and productive bot detection system can detect many forms of botnets with various behavioral features. The proposed method can also be used on wide scale datasets and is resistant to zero-day threats. This section explains how we used an experiment to investigate the influence of deep learning representation plans on intrusion detection. We examine the structure of the autoencoder model as well as the magnitude of its latent dimension. The procedure to construct the best autoencoder model is designated in Figure 6.3.

6.5.3.1 Model Design

For irregularity-based interruption discovery schemes, the diverse autoencoder models are employed as unverified mechanism knowledge procedures. We trial with various perfect shapes to see if the design decisions made by an autoencoder model affect intrusion detection performance. We spotlight on two important plan features of autoencoder models in this paper: model structure and latent size. We change depth and width of the model to see how capacity and complexity affect detection performance.

We examine numerous mixtures of alternative perfect constructions and dormant sizes using an autoencoder model built, totally associated unseen layers with the ReLU beginning role and a yield layer with the sigmoidal initiation role. Three model structure configurations were evaluated for assessment. Each model structure arrangement as shown in the table is represented as a (Depth, Size). The number of completely linked out of sight layers in the model is indicated by depth.

We compare the (5, 32) perfect to the (5, 64) model, which have similar quantity of hidden pieces in the amount of trainable limits, to see if perfect volume has a bearing on IDS presentation. We compare the (5, 64) and (7, 64) models to see how model complexity affects performance. They have equal numbers of trainable limits, but differing numbers of hidden layers. Table 6.3 demonstrates the proposed model structure of autoencoder. We performed our investigation in (5, 32), (5, 64), and (7, 64) models.

We examine the potential latent size for a piece perfect construction, which ranges from 1–1. We vary the dormant coating dimensions reaching from 1 to 15 in the (5, 32) and (7, 64)

TABLE 6.3

Proposed Model Structure

Prototypical Construction (Deepness, Dimension)	No. of Nerve Cells in Every Layer
(_5, 32_)	_Input_32_16_latent layer_16_32_output
(_5, 64_)	_Input_64_32_latent layer_32_64_output
(_7, 64_)	_Input_64_32_16_latent layer_16_32_64_output_

models, for example. The latent sizes of 1–31 are tested in the (5, 64) model. The Adam optimizer is used to train these models to minimize the reconstruction loss, using a learning degree of one 104 and a weight decay value of 1,105. There were no further regularization techniques utilized.

The procedure contains four stages: preprocessing, exercise, model selection, selection of threshold and proof.

- **Preprocessing:** From the exercise dataset, first filter out attack, leaving only usual examples to train the auto encoder with, further by means of the min-max standardization.
- **Training:** With the L2 norm as the reconstruction loss, closely as likely to the training set's usual record designs are found. The autoencoder model that was trained with normal logs should theoretically be able to get well. As a result, the recovered output will almost certainly differ significantly from the original input.
- **Model Selection:** The AUC score allows us to evaluate the model's general presentation because of change in model weights.
- **Selection of the Threshold:** The verge on the renovation coldness connecting a contribution and its production divide the normal and aberrant classes; the value chosen has a significant influence on ML-NIDS performance.
- **Validation:** The validation is used to evaluate the model's performance.

6.5.3.2 Experiment Setup

For our investigation, we utilized Google Colaboratory to quickly assess detection accuracy. Google supplies us with a powerful platform that includes a high-performance CPU and GPU to accelerate the process. As a result, the overall time necessary to calculate the findings is reduced. For easy maintenance, all of the datasets are saved in Google Drive. All of the Google Colaboratory's inputs come from Google Drive. Two deep learning algorithms were used to compare the findings during the assessment phase. The characteristics picked by the chi-square feature selection approach were used to train each model. The hyper-parameters in each layer of the neural network were tweaked to fine-tune the deep learning models.

6.5.4 Performance Metrics and Result Evaluation

We used the N-BaIoT dataset, which was divided into several sets based on the kind of IoT device. For each device type, we present the average of five runs, each of which is qualified for 50 epochs with a lot size of 1,024. Table 6.4 shows the standard show metrics for five runs of each model setting for the specified devices.

TABLE 6.4

Evaluation Results of the Proposed Model and Existing Model with N-BaIoT Dataset

Danmini Doorbell

Representation Model	Latent Size (LS)	Threshold (Th)	Accuracy (Acc)	True-Positive Rate (TPR)	False-Positive Rate (FPR)	MCC	F1	AUC
CNN	10	0.978	1.000	1.000	0.050	0.986	1.000	0.909
RNN-LSTM	15	0.988	1.000	1.000	0.058	0.989	1.000	0.912
(5, 32)	11	0.987	1.000	1.000	0.006	0.996	1.000	0.999
(5, 64)	19	1.1.06	1.000	1.000	0.006	0.995	1.000	0.999
(7, 64)	8	1.023	1.000	1.000	0.006	0.995	1.000	0.999

Baby Monitor (Philips)

Model	Hidden Size (LS)	Threshold (Th)	Accuracy	TPR	FPR	MCC	F1	AUC
CNN	14	2.987	0.910	0.993	0.014	0.902	0.998	0.934
RNN-LSTM	23	2.989	0.923	0.994	0.011	0.930	0.998	0.976
(5, 32)	10	3.130	0.995	0.998	0.013	0.986	0.998	0.999
(5, 64)	25	3.270	0.997	0.998	0.010	0.989	0.998	0.999
(7, 64)	8	3.322	0.996	0.999	0.011	0.985	0.998	0.999

Simple Home XCS7–1003-WHT Security Camera

Model	Hidden Size (LS)	Threshold (Th)	Accuracy	TPR	FPR	MCC	F1	AUC
CNN	18	0.345	0.890	0.999	0.045	0.912	0.995	0.989
RNN-LSTM	19	0.879	0.965	0.999	0.065	0.963	0.996	0.090
(5, 32)	11	0.438	0.996	0.999	0.074	0.951	0.996	0.990
(5, 64)	20	1.040	0.996	0.999	0.051	0.964	0.997	0.998
(7, 64)	12	0.846	0.995	0.999	0.058	0.956	0.997	0.996

Ecobee – Thermostat

Model	Hidden Size (LS)	Threshold (Th)	Accuracy	TPR	FPR	MCC	F1	AUC
CNN	15	0.779	0.898	0.933	0.056	0.89	0.992	0.991
RNN-LSTM	18	0.438	0.954	0.977	0.061	0.911	0.994	0.995
(5, 32)	13	0.802	0.997	0.999	0.067	0.928	0.999	0.997
(5, 64)	4	0.408	0.999	1.000	0.064	0.959	0.999	0.999
(7, 64)	4	0.366	0.999	1.000	0.079	0.952	0.999	0.999

Security Cam (XCS7–1002-WHT)

Model	Hidden Size (LS)	Threshold (Th)	Accuracy	TPR	FPR	MCC	F1	AUC
CNN	18	0.345	0.890	0.999	0.045	0.912	0.995	0.989
RNN-LSTM	19	0.879	0.965	0.999	0.065	0.963	0.996	0.090
(5, 32)	11	0.438	0.996	0.999	0.074	0.951	0.996	0.990
(5, 64)	20	1.040	0.996	0.999	0.051	0.964	0.997	0.998
(7, 64)	12	0.846	0.995	0.999	0.058	0.956	0.997	0.996

(Continued)

TABLE 6.4 (*Continued*)

Evaluation Results of the Proposed Model and Existing Model with N-BaIoT Dataset

Provision_PT-737E Security Camera

Model	Hidden Size (LS)	Threshold (Th)	Accuracy	TPR	FPR	MCC	F1	AUC
CNN	18	0.345	0.889	0.999	0.045	0.912	0.995	0.989
RNN-LSTM	19	0.879	0.965	0.999	0.065	0.963	0.996	0.090
(5, 32)	10	0.854	0.991	0.994	0.050	0.937	0.995	0.992
(5, 64)	29	1.338	0.996	0.999	0.026	0.976	0.998	0.999
(7, 64)	9	1.103	0.995	0.998	0.049	0.967	0.998	0.999

Provision_PT-838 Security Camera

Model	Hidden Size (LS)	Threshold (Th)	Accuracy	TPR	FPR	MCC	F1	AUC
CNN	18	0.346	0.890	0.999	0.045	0.912	0.995	0.989
RNN-LSTM	19	0.879	0.965	0.999	0.065	0.963	0.996	0.090
(5, 32)	10	0.954	0.991	0.994	0.050	0.937	0.995	0.992
(5, 64)	29	1.338	0.996	0.999	0.026	0.976	0.998	0.999
(7, 64)	9	1.103	0.995	0.998	0.049	0.967	0.998	0.999

Security Camera (XCS7–1002-WHT)

Model	Hidden Size (LS)	Threshold (Th)	Accuracy	TPR	FPR	MCC	F1	AUC
CNN	18	0.345	0.890	0.999	0.045	0.912	0.995	0.989
RNN-LSTM	19	0.879	0.965	0.999	0.065	0.963	0.996	0.090
(5, 32)	10	0.854	0.991	0.994	0.050	0.937	0.995	0.992
(5, 64)	29	1.338	0.996	0.999	0.026	0.976	0.998	0.999
(7, 64)	9	1.103	0.995	0.998	0.049	0.967	0.998	0.999

Samsung_SNH_1011_N Webcam

Model	Hidden Size (LS)	Threshold (Th)	Accuracy	TPR	FPR	MCC	F1	AUC
CNN	11	2.978	0.910	0.993	0.014	0.920	0.998	0.934
RNN-LSTM	22	2.998	0.995	0.994	0.014	0.940	0.998	0.976
(5, 32)	10	3.230	0.995	0.998	0.016	0.987	0.998	0.999
(5, 64)	25	3.470	0.997	0.998	0.017	0.989	0.998	0.999
(7, 64)	8	3.722	0.996	0.999	0.019	0.989	0.998	0.999

6.5.4.1 Evaluation Metrics

For comparing alternative model configurations, conventional measures such as correctness, true optimistic rate, and false-positive rate are used. The true-positive rate (TPR) is a metric that measures how well an attack is detected. The following equations are used to determine TPR, FPR, accuracy, and F1.

$$\text{True\_Positive\_Rate} = \frac{TP}{TP + FN} \tag{6.1}$$

$$\text{False\_Positive\_Rate} = \frac{FP}{FP + TN} \tag{6.2}$$

$$Accuracy = \frac{TP + TN}{TP + FP + TN + FN} \tag{6.3}$$

$$F1 = \frac{TP}{TP + 1/2(FFP + FN)} \tag{6.4}$$

We similarly use the Matthews correlation coefficient (MCC) as the key statistic to demonstrate the model's optimal presentation when the threshold is fixed and utilized to determine the score, which runs from 1 to (1,1). A textbook guess, whereas one shows that all forecasts are incorrect.

6.5.4.2 Result Evaluation

As a result, even when there is a class imbalance, MCC may effectively reflect the model's performance and is used to assess classification performance.

6.5.4.3 Model Structure and Performance

Our main findings are summarized and discussed in this section. We begin by examining the link between perfect construction and presentation in relations of perfect facility and strength. Crossways all of the standard datasets, the perfect (5, 64) – which has five unseen layers and 64 neurons. It says that model competence has a collision on presentation models with greater capacities performing better. However, we found no link between model depth and performance. Table 4 demonstrates the results of the existing model and proposed model with N-BaIoT dataset.

Second, when the latent size changes, we evaluate the IDS performance. We found that when the underlying dimension rises up to a positive point, the total intrusion detection performance improves and then declines. This implies that the representation structure is stable and adjusting the latent dimension may considerably increase performance. Figure 6.4 demonstrates the result of an autoencoder model with (7, 64) layers for nine devices of the N-BaIoT dataset, in which the green color indicates nominal value and the red color indicates anomaly. Here, anomalies can be predicted as attacks.

6.6 Conclusions

Unsupervised deep learning techniques such as autoencoder have been investigated extensively for network intrusion detection because they can quickly identify zero-day attacks and considerably reduce labeling labor. Although autoencoder representation seat been demonstrated to be operative in detection impositions, choosing the best perfect construction to get the best detection performance takes time and effort. We used an autoencoder model with several perfect shapes to construct and evaluate the model. N-BaIoT empirical findings show IDS performance is influenced by the size of the model and the size of the bottleneck layer, according to our findings. A proposed autoencoder model, in particular, performs better and is more stable as the model size becomes larger.

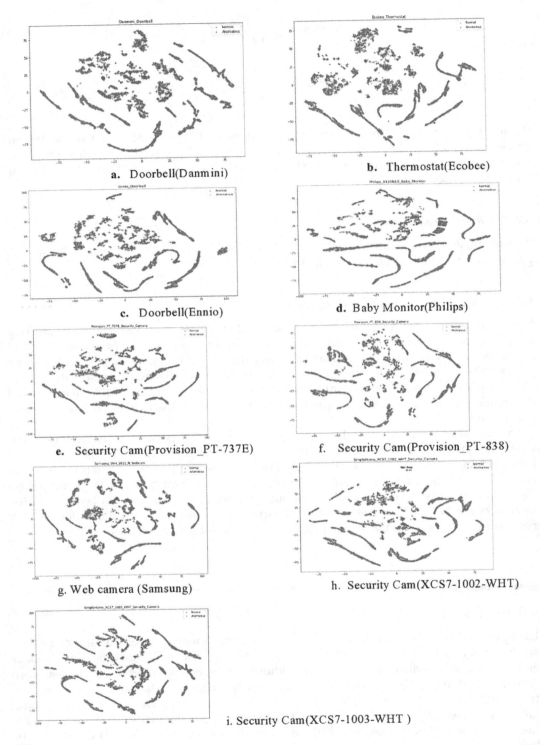

a. Doorbell(Danmini)

b. Thermostat(Ecobee)

c. Doorbell(Ennio)

d. Baby Monitor(Philips)

e. Security Cam(Provision_PT-737E)

f. Security Cam(Provision_PT-838)

g. Web camera (Samsung)

h. Security Cam(XCS7-1002-WHT)

i. Security Cam(XCS7-1003-WHT)

FIGURE 6.4
Normal and anomalies in each device (a–i).

Experiments with a restricted model ability, deepness, and latent dimension are the focus of this collection of combinations yielded our findings. The impact of depth, in particular, was not completely proven and will need to be examined further with a greater autoencoder perfect. Because the feature vector determines the autoencoder's input layer, obtaining datasets with big input feature vectors is essential to develop a large model. The size of an autoencoder's hidden layers decreases until it reaches the bottleneck layer. Future research might broaden our scope to include other areas of model design as well as a larger range of setups and datasets. We will also investigate approaches without scanning all potential ranges and determine the appropriate latent size.

References

[1] Nazir, A. and Khan, R.A., Network intrusion detection: Taxonomy and machine learning applications, *Machine Intelligence and Big Data Analytics for Cybersecurity Applications (Studies in Computational Intelligence)*, 919, 3–28, 2021.

[2] Ramakrishnan, V., Chenniappan, P., Dhanaraj, R. K., Hsu, Bootstrap aggregative mean shift clustering for big data anti-pattern detection analytics in 5G/6G communication networks, *Computers & Electrical Engineering*, 95, 1–17, 2021.

[3] Bhaskar, K. B., Prasanth, A., Saranya, P., An energy-efficient blockchain approach for secure communication in IoT-enabled electric vehicles, *International Journal of Communication System*, 35, 1–25, 2022.

[4] Rashad, J. M., and Samuel, B. O., An exploratory study on the use of internet_of_medical_ things (IoMT) In the Healthcare Industry and their Associated Cybersecurity Risks, *Int'l Conf. Internet Computing and Internet of Things (ICOMP'19)*, 1–6, 2021.

[5] Lavanya, S., Prasanth, A., Jayachitra, S., A Tuned classification approach for efficient heterogeneous fault diagnosis in IoT-enabled WSN applications, *Measurement*, 183, 1–22, 2021.

[6] Maria, P., et al., A survey on security threats and countermeasures in internet of medical things (IoMT), *Transactions on Emerging Telecommunications Technologies*, 1–15, 2020.

[7] Neshenko, N. et al., Demystifying IoT security: An exhaustive survey on IoT vulnerabilities and a first empirical look on internet-scale IoT exploitations. *IEEE Communications Surveys and Tutorials*, 21, 2702–2733, 2019.

[8] Brun, O., Yin, Y., Gelenbe, E. Deep learning with dense random neural network for detecting attacks against IoT-connected home environments. *Procedia Computer Science*, 134, 458–463, 2018.

[9] Doshi, R., Apthorpe, N., Feamster, N. Machine learning DDoS detection for consumer internet of things devices. In *Proceedings of the 2018 IEEE Security and Privacy Workshops (SPW), San Francisco, CA, USA, 24–24 May*, pp. 29–35, 2018.

[10] Challoner, A., Gheorghe, H.P., Intelligent sensing technology, smart healthcare services, and ear internet of medical things-based diagnosis. *American Journal of Medical Research*, 6(1), 13–18, 2019.

[11] George, W.C., Michael, V.D, Todd, R.A. Cybersecurity issues in robotics. In *Cognitive and Computational Aspects of Situation Management (CogSIMA), 2017 IEEE Conference on, IEEE.* pp. 1–5, 2017.

[12] Deogirikar, J., Vidhate, A. Security attacks in IoT: A survey. In *2017 International Conference on I-SMAC (IoT in Social, Mobile, Analytics and Cloud) (I-SMAC)*, pp. 32–37, 2017.

[13] Mandy, D., et al. An overview of steganography techniques applied to the protection of biometric data. *Multimedia Tools and Applications*, 77(13), 17333–17373, 2018.

[14] Fei, J., et al. Artificial intelligence in healthcare: Past, present and future. *Stroke and Vascular Neurology*, 2(4), 230–243, 2017.

[15] Meidan, Y. et al. IoT: A machine learning approach for IoT device identification based on network trace analysis. In *Proceedings of the Symposium on Applied Computing, Marrakech, Morocco, 4–6 April*, pp. 506–509, 2017.

[16] Hatice, C.K., Aydogan, O., Wearable and implantable sensors for biomedical applications. *Annual Review of Analytical Chemistry*, 11, 127–146, 2018.

[17] He, D., Chan, S. Guizani, M., Drone-assisted public safety networks: The security aspect. *IEEE Communications Magazine*, 55(8), 218–223, 2017.

[18] Mandy, D., et al. An overview of steganography techniques applied to the protection of biometric data. *Multimedia Tools and Applications*, 77(13), 17333–17373, 2018.

[19] Dowling, S., Schukat, M., Melvin, H. A zigbee honeypot to assess IoT cyberattack behaviour. In *Signals and Systems Conference (ISSC), 2017 28th Irish*, pp. 1–6, IEEE, 2017.

[20] Dhiviya, S., Malathy, S., Kumar, D. R. Internet of things (IoT) elements, trends and applications. *Journal of Computational and Theoretical Nanoscience*, 15(5), 1639–1643, 2018.

[21] Duy La, Q., Quek, TQS, Lee, J. A game theoretic model for enabling honeypots in IoT networks. In *Communications (ICC), 2016 IEEE International Conference on*, pp. 1–6, IEEE, 2016.

[22] Brendan, W.M. et al. Augmented reality in emergency medicine: A scoping review. *Journal of Medical Internet Research*, 21(4), e12368, 2019.

[23] Liu, X. et al. Ingestible hydrogel device. *Nature Communications*, 10, 2019.

[24] Noura, H., Chehab, A., Couturier, R. Lightweight dynamic key-dependent and flexible cipher scheme for IoT devices. In *2019 IEEE Wireless Communications and Networking Conference (WCNC)*, pp. 1–8, IEEE, 2019.

[25] Colin, J.M., Raul, N.U. Advances in virtual and augmented reality – exploring the role in health-care education. *Journal of Radiology Nursing*, 2019.

[26] Sophie, M., Anne, C., Deborah, L., The effect of telehealth versus usual care for home-care patients with long-term conditions: A systematic review, meta-analysis and qualitative synthesis. *Journal of Telemedicine and Telecare*, 1357633X19862956, 2019.

[27] Emma, M., et al. Assessing medical device vulnerabilities on the internet of things. In *2017 IEEE International Conference on Intelligence and Security Informatics (ISI)*, pp. 176–178, IEEE, 2017.

[28] Melki, R., et al. An efficient of DM-based encryption scheme using a dynamic key approach. *IEEE Internet of Things Journal*, 1–1, 2018.

[29] Masdari, M., Khezri, H. A survey and taxonomy of the fuzzy signature-based intrusion detection systems. *Applied Soft Computing*, 92, 106301–106320, 2020.

[30] Zhao, D., et al. Botnet detection based on traffic behavior analysis and flow intervals. *Computers and Security*, 39, 2–16, 2013.

[31] Erlacher, F., Dressler, F. On high-speed low based intrusion detection using snort-compatible signatures. *IEEE Transactions on Dependable and Secure Computing*, 14, 2020.

[32] Raheja, S. G., et al. Rule-based approach for botnet behavior analysis. In *Intelligent Data Analytics for Terror Threat Prediction: Architectures, Methodologies, Techniques and Applications*. Wiley, pp. 161–179, 2021.

7

Using Self-Organizing Map to Find Cardiac Risk Based on Body Mass Index

Pon Bharathi A and Allan J. Wilson
Amrita College of Engineering and Technology

S. Veluchamy
Sri Venkateswara College of Engineering

S. Swathi
Sri Sairam Institute of Technology

CONTENTS

7.1 Introduction

Even though the latter is more convenient to integrate into some specialized applications, a biometric system based on physiological variables is more reliable than the one based on behavioral variables. It is widely assumed that biometrics will become a significant part of detection systems as the cost of biometric sensors continues to fall and the public becomes more aware of biometrics' strengths and limitations [1]. When facial or iris recognition is used to control access to restricted areas within a facility, it can generate a smooth entry experience. The monitoring and statistical analysis of people based on their unique attributes is called biometrics. Identification and access control, as well as other methods of identifying people who have been followed, are all viable uses of the technology. Fingerprint recognition is the most commonly used biometric for accessing buildings, so it provides an extra layer of protection in building management [2].

Fundamental research into these modalities has reached a pinnacle, but certain key characteristics, including multidimensional space and multiband hyperspectral space, are still being researched to optimize their utility. Researchers are also looking at biosignals such as electrocardiography (ECG), electroencephalography (EEG), and electromyography (EMG) for biometric system efficiency and long-term reliability [3]. Biometric recognition

DOI: 10.1201/9781003256243-7

using a person's heartbeat has been under study for more than a decade already. Heart-based biometrics' most effective clinical test, the ECG, was thoroughly explored. Researchers have been concentrating on phonocardiogram (PCG) and, more recently, photoplethysmogram (PPG) signals from the heart (heart sounds) in addition to ECG (heart pulses).

Heart rate variability (HRV) is the natural formation of the heart that has been the focus of medical research during the past two decades [4]. Biometrics, on the other hand, is still in its early stages of development. Several attempts have been made to classify HRV data in order to identify disease abnormalities. In fact, only two early attempts at HRV recognition have been published in the literature. There is no known attempt at biometric recognition; however, HRV-based human identification by Irvine et al. is the only known effort for biometric recognition, even if its methodology and results are dubious due to a lack of data. Using HRV in biometric identification systems and using heart-based biometrics are two of the primary themes of this chapter. The database template is compared to a feature set chosen from the data. It is possible to employ a biometric system for identification or verification, depending on application context. This mode is used to validate a person's identity by comparing biometric data collected to the system database's stored biometric templates.

There are many different ways to implement a biometric system, but the most common method is to collect biometric data, extract features, and compare them to database templates [5]. A biometric system may be used in either or identification mode, depending on the environment in which it is implemented. As part of a person's identity verification, a biometric database is compared to her own personal biometric template(s). The system uses one-to-many comparison to determine a person's identification (or fails to do so if that individual is not already registered in the system database), without having to ask "Whose biometric data are this?" When a person denies being who he/she claims to be, the system must be able to determine whether or not he/she is being truthful. Negative recognition aims to prevent a single person from taking on multiple identities [6]. Additionally, it is possible to take advantage of the convenience of positive reinforcement. (It is not necessary for the user to declare his or her identity.)

The system recognizes a user in identification mode by scanning the database for a match between all of the user's templates. To validate an individual's identity, the system uses a one-to-many comparison since there is no requirement for the subject to claim identification (e.g., "Whose biometric data are this?"). (If the subject is not registered in the system database, the operation fails.) The identification is necessary for negative recognition applications in order to verify the individual's (implicit or explicit) assertions of identity. In order to prevent one individual from establishing many identities, negative recognition is used. IDs may also be used in positive recognition for the benefit of ease of access. (It is not necessary for the user to declare his or her identity.)

In recent years, biometrics (such as fingerprints or facial recognition) have largely replaced the traditional password and PIN as a method of user identification, particularly in personal and mobile devices. One of the advantages of cardiac biometrics is aliveness detection, which gives a strong challenge to spoofing attempts. To date, a number of authentication methods based primarily on the ECG have looked excellent. However, there is currently a scarcity of research into other cardiac signals [7].

An individual's overall health and the likelihood of getting certain illnesses, such as heart disease, may be determined by analyzing a person's biometric data. Knowing and improving one's biometric scores may have a profound effect on one's health. Regular biometric screenings measure things such as a person's weight, height, waist circumference, blood pressure, blood glucose, cholesterol, and triglycerides [8]. Knowing and understanding one's biometric data is critical for people who are worried about cardiovascular disease. Doctors weigh us, and many of us do not like the experience. The likelihood of a person acquiring heart disease or suffering a heart attack, or any of a number of other diseases and illnesses, including particular disorders, may be estimated using the body mass index (BMI).

In order to assess whether a person is underweight, of normal weight, or overweight, the BMI is used to quantify the individual's weight. There are several health hazards associated with being overweight or obese. A person's BMI does not take into account age, sex, ethnicity, or muscular mass when calculating their body fat percentage. The BMI, which is based only on weight and height, determines these levels. Obesity, on the other hand, is affected by age and gender. For example, if two people's ages are 88 and 22 and their weights are the same, their BMI will be the same. Both people, however, have varying obesity levels. There is no mathematical formula that we are aware of that explains and/or determines the obesity gap. Machine learning algorithms may be used to create reliable obesity estimates in a wide range of situations [9].

There are a number of neural network models that are often employed without supervision, including the self-organizing map (SOM). Clustering, dimension reduction, and feature identification have all been accomplished using SOM [10]. One of the first proponents of SOM was Professor Kohonen. As a result, the SOM is also known as the Kohonen map. Machine condition monitoring, fault identification, satellite remote sensing, and robot control are just a few of the real-world uses.

SOM's architecture is extremely similar to that of a neural network; however, it is far simpler than that of an artificial neural network (ANN) [11]. SOM has two layers: an input layer and an output layer (feature map). SOM begins feature mapping by constructing a weight vector. Weights have an entirely different meaning in SOM compared to ANN. When modeling a neural network, the activation function is used to produce a linear combination of weights and input values to produce an output for each of the neurons in the design. The activation function, on the other hand, is not used by SOM. Weights are used as the neuron's features in the design. The majority of the weights are generated at random. The weight vector is used as a neuron characteristic to push each row (observation) of data into an imagined space where each row works as a point. This is the point at which the SOM begins. A search for the nearest points is conducted once each item of data has an imaginary point in the input space [12].

The purpose of this research is to figure out how likely heart disease and heart attacks are. This study aims to create a deep learning-based SOM technique for estimating BMI and detecting ECGs. The proposed BMI technique is tested using an ECG dataset with changing HRV data. HRV is a statistical analysis of the ECG that can disclose information about a person's response to external or internal physical or emotional activity. HRV is a measure that assesses the impact of stress on the human body. HRV values differ from person to person. Individual HRV baseline readings can help overcome this variability.

7.2 Literature Survey

This work investigates the methods to determine the risk of cardiac illness, such as a heart attack, as well as the detection of certain tumors. This approach is generally used to compute the BMI based on the weight and height of the folks for determining the risk using a deep learning-based SOM approach.

Akhter et al. examined how HRV is employed in biometric recognition technologies and described the present state of heart-based biometrics [13]. Although basic research into these modalities has reached a pinnacle, crucial components such as multidimensional space and multiband hyperspectral space are still being investigated in order to improve their utility. Biosignals are being investigated to improve the efficiency and resilience of biometric systems, including ECG, EEG, and EMG. The human heart has been studied as a potential biometric recognition solution for more than a decade. In terms of cardiac biometrics, the most dependable clinical activity, ECG, has been the most extensively studied. Researchers have been concentrating on phonocardiogram (PCG) and, more recently, photoplethysmogram (PPG) signals from the heart (heart sounds) in addition to ECG (heart pulses). This study focuses on heart-based biometrics as well as the practicality of integrating HRV in biometric identification systems.

Wang et al. are a group of researchers who have worked on a number of different projects [14]. The goal of this research is to come up with a new way to identify ECG signals that last shorter than a minute. Re-sampling of the pulse to remove the influence of HR fluctuation is advocated in this study. For the purpose of examining any differences between patients, the principal component analysis network (PCANet) is utilized. To test the recommended strategy, researchers used a public ECG-ID database with a variety of HR data from various individuals.

The ECG signal is a record of the heart's pulsating rhythms. For identification purposes, the ECG signal provides a great deal of information about the patient, and a heartbeat is one phase of this signal. Because the amplitude and duration of the beats alter so little, the visual interpretation of the beats is difficult. In ECG identification, pattern recognition algorithms are used because they are reliable, fast, and objective. As described above, the goal of this research is to create a system that can recognize short-term HR ECG data. Using ECG data, the suggested approach may correctly identify the patient 94.4% of the time with just five heartbeats, according to testing findings.

It is important to take into consideration individual cholesterol levels in order to analyze how effectively two summary estimates predict total CHD when compared to the current data [15]. Instead of focusing on the relevance of individual lipoprotein cholesterols, this presentation aims to examine the summary estimations to see whether any critical data that might help define CHD risk are missing. We can already see this in the ratio of low-density lipoprotein cholesterol to bigger lipoprotein cholesterol, but it will become much more obvious in future research as our capacity to analyze lipoprotein cholesterol dynamics develops. This research uses the term "lipoprotein cholesterol" to characterize the cholesterol levels in distinct lipoprotein classes. The key topic addressed in this study is how well summary cholesterol estimates, when used alone or in combination with information on individual cholesterol levels, predict the development of CHD.

In [16], a number of variables linked to extremely high risk in women are examined. Diabetes and a low amount of high-density lipoprotein cholesterol (HDL-C) are the primary components of this constellation. Obesity may be a significant link between these two high-risk characteristics. The Framingham Study analyzed blood lipids using the Fredrickson and Levy procedure after an overnight fast and then monitored participants for the development of cardiovascular disease. Subsequent to controlling for the impacts of HDL-C and diabetes, fasting fatty substances were not. In contrast to men, women with low degrees of HDL-C had a higher danger of coronary illness in the setting of diabetes.

In [17], a short-time frequency biometric ECG analysis with robust feature selection is shown. The suggested method takes into consideration heartbeats over a period of time. An equivalent error rate (EER) of 5.58% was found while using heartbeats from different days to train the proposed approach, with a 74.9% rate of identification and a 93.5% rate of recognition for rank 15. To attain a 0.37% EER and a 99% identification accuracy, we need to capture both training and test heartbeats on the same day.

Huang et al. explained how to leverage deep learning (DL) techniques, notably convolutional networks, to obtain usable representation for cardiac biometric identification in this research [18]. Considering two separate channels: the raw pulse signal and the heartbeat spectrogram, we focus on constructing training data for heart biometrics. They also go through heartbeat data augmentation algorithms, which are crucial for DL systems to generalize.

The echocardiography is a reliable method for evaluating a fetus at risk of congenital heart disease during pregnancy [19]. Pregnancies with a variety of risk factors should be evaluated to identify afflicted fetuses, and the results should be used to develop perinatal and neonatal care regimens. The first phase of each fetal echocardiography was to get a four-chamber image of the heart. The four-chamber view may be a useful tool for detecting congenital heart disease, and it should be included in all normal obstetric ultrasound examinations.

On the other hand, the possibility of suppressing Notch signaling has been presented in order to prevent infection with SARS-CoV-2 and to reduce the course of coronavirus disease-related heart and lung disease [20]. Several molecular pathways that mediate viral infections have been linked to Notch signaling. SARS-CoV-2 infection and the advancement of COVID-19-related heart and lung illness might be prevented by blocking Notch signaling, according to this study.

Widiyaningtyas et al. discussed CHD [21]. They mainly defined the risk of heart disease and death rate and how data are taken and tested using the SOM method; age, gender, and type of chest pain were the major considerations. Blood pressure and blood sugar levels were evaluated, as well as the type of chest pain. The latest results identify an individual as having heart disease or not having heart disease depending on the value of each trait. As a result, the accuracy is 70%.

Chatterjee et al. described the CVD and mainly how these risks occur based on the health issues of the person [22], such as obesity and overweight. Thirty percent of global fatalities are predicted to be caused by lifestyle diseases, which are preventable if risk factors and behavioral interventions are identified and implemented, according to the WHO. Health behavior improvement should be addressed in order to avert life-threatening outcomes. This study's main purpose was to analyze numerous machine learning algorithms and their implementations based on publicly available sample health data related to lifestyle illnesses including obesity and cardiovascular disease, rather than to create a risk prediction model.

Steven et al. discussed research into the use of pulse data as a biometric for determining a person's identity [23]. A variety of approaches for acquiring heartbeat signatures, as well as processing methods, have been presented. Individual variants, environmental variants, and sensor variants are described in this work as three essential aspects that influence performance, which they use to address the biometric identity and verification challenge. All three of these factors have an impact on the signal's ability to be collected and analyzed, used for individual identification or verification, and disseminated.

Heartbeat measurements are acquired, analyzed, and exploited as part of person identification based on cardiac performance. The obstacles are divided into three basic categories, each of which is highly dependent on the application: the collection environment, target variability, and sensor efficacy. In order to record a heartbeat, sensors, environments, and targets must all work together. The heartbeat detection problem will be treated as a signal processing problem in order to uncover the operational issues that need to be addressed.

The unsupervised neural networks were used in this research [24] to examine the spread of the coronavirus throughout the globe. This allowed us to put countries together depending on the number of coronavirus cases they had, enabling us to analyze if countries were responding similarly and hence may benefit from utilizing comparable strategies in the fight against the virus's spread.

The most common cause of death in the USA is coronary artery disease [25]. High blood pressure, obesity, and diabetes are just a few of the risk factors for cardiovascular disease. All of these CVD risk variables have improved as a result of intensive therapeutic lifestyle modification programs (ITLMPs). BMI tracking could be a useful proxy variable for detecting weight-loss program adherence and forecasting the deterioration of baseline health outcomes after undergoing a rigorous lifestyle change program.

By analyzing neuroimaging data, deep neural network models may detect anomalies in the brain and identify those at risk of cognitive decline and neurodegenerative illness [26]. Using structural brain imaging and DL for the first time, they investigated whether BMI could be predicted using these techniques. Individual BMI may successfully be predicted with the use of a single structural MRI brain scan, as well as age and gender information, according to the researchers.

In localization maps created for the CNN, caudate nucleus and the amygdala were found as brain areas that greatly contributed to BMI prediction [27]. The CNN-based visualization approach adds to the data demonstrating the relationship between brain anatomy and BMI when compared to the findings of a conventional automated brain segmentation method. Based on our findings, it is possible to use DL to predict BMI from structural brain scans and learn more about the relationship between brain morphological diversity and individual differences in weight variation. They also open up new research pathways into the clinical relevance of brain-predicted BMI in the future.

7.3 Methodology

Artificial intelligence (AI) techniques based on human learning are called deep learning. In the field of data science, which encompasses statistics and predictive modeling, deep learning is an essential component. To speed up and simplify the process of collecting, analyzing, and interpreting vast volumes of data, deep learning is an invaluable tool for data scientists. Among the many learning approaches used in AI and machine learning, DL models stand out as a particularly useful one. Big data disciplines are becoming more interested in picture and voice analysis, which is a result of recent advances in the field. The negative is that the mathematics and computer technology of deep learning models is very difficult to understand, particularly for researchers from other disciplines. "At last, we offer an outline of profound learning draws near, including deep feedforward neural networks, convolutional neural networks, deep belief networks (DBNs), autoencoders (AEs), and long short-term memory (LSTM) networks in this chapter."

Although DL has been around for a while, it is currently receiving a lot of attention. Multilayer neural networks with several hidden units are examples of the kind of big prediction models that may be learned using DL. DL is a collection of approaches rather than a single method. Additionally, deep learning has been utilized to solve a range of problems in a variety of industries. In the MNIST dataset, for example, a deep learning algorithm achieved a new record for handwritten digit classification with an error rate of 0.21%.

Image recognition, speech recognition, natural language comprehension, acoustic modeling, and computational biology are some of other applications.

DL is a type of machine learning that takes an input X and uses it to predict an output Y. Given a large dataset of source and target pairs, a deep learning system will try to close the gap between its prediction and expected output. By attempting to comprehend the association/pattern between given inputs and outputs, a deep learning model can generalize to inputs it hasn't seen before. There are five stages in all in this paper: (a) database, (b) initialization, (c) SOM process, (d) validation, and (e) analysis.

a. Database

This database contains data that are used to calculate BMI. Adults' height, weight, and gender are listed in the data table. The BMI value of each male and female person's fat level and cardiac risks will be determined using the SOM method based on their gender's height and weight. Table 7.1 illustrates the dataset of BMI.

b. Data Preprocessing

Here, collect the data for feeding into the Keras model during the preprocessing stage. A first idea is to reduce all null values from the database. Then, to transform category parameters to numerical variables, we'll utilize one-hot encoding. The data that neural networks deal with are numerical rather than categorical. It also divides the data into two groups: training and testing. Finally, we'll scale/standardize the data so that they fall between −1 and 1. This standardization aids the model's training and allows it to converge more quickly.

Children, their families, and the physical environments they grew up in all contributed to the collection of data on a wide range of features, including BMI, height, weight, obesity flags, and other markers of a person's socioeconomic situation. As can be seen in Figure 7.1, obesity flags for 14-year-olds were classified as either "normal" or "at risk," depending on the findings. In Figure 7.1, the machine learning system employed BMI data from the ages of 3, 5, 7, and 11 to set data levels, while obesity flags for the ages of 14 were used as outcomes in various ways, such as normal and at risk (overweight).

The following elements describe obesity warning symptoms among 14-year-olds:

- At the time of the survey, what was your age?
- Height in meters and weight in kilos
- Gender.

TABLE 7.1

Dataset of BMI

S. No.	Gender	Height	Weight
0	Male	174	96
1	Male	189	87
2	Female	185	110
3	Female	195	104
4	Male	149	61
5	Male	189	104
6	Male	147	92
7	Male	154	111
8	Male	174	90

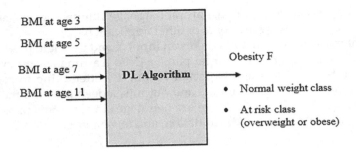

FIGURE 7.1
BMI measure.

TABLE 7.2

BMI Range

Below 18.5	Underweight
18.5–24.9	Normal or healthy weight
25.0–29.9	Overweight
25.0–29.9	Overweight

In Eq. (7.1), your height and weight are used to calculate your BMI, which may help establish whether or not you are a healthy weight. Weight in kilograms is divided by the squared height in meters to get the BMI of a person. If you have a BMI of 25, your weight is 25 km/m².

$$BMI = \frac{Weight(kg)}{Height(m)^2} \tag{7.1}$$

7.3.1 BMI Scales

The optimum BMI range for most people is between 18.5 and 24.9. Children and adolescents between the ages of 2 and 18 are estimated to have a BMI between 18.5 and 24.9. The BMI ranges are shown in Table 7.2.

Hyperlipidemia, increased insulin, and high blood pressure are all linked to an individual's BMI-for-age. At least one biochemical or clinical risk factor for cardiovascular disease, such as those mentioned above, was found in more than 60% of overweight children aged 5–10, whereas 20% of these children had two or more risk factors.

In middle age, lipid and lipoprotein levels, as well as blood pressure, are linked to BMI-for-age during pubescence. Children's risk factors can develop into chronic diseases in adults. In teenagers, BMI-for-age has a substantial relationship with subcutaneous and total body fatness. In more than 95% of the population, overweight risk ranges from the 85th to the 95th percentile. Underweight is the 5th percentile. Those with a BMI of 95% or above are termed obese, whereas those with a BMI between 85% and 95% are not considered obese, but at danger of becoming so. When a child's BMI-for-age is over or equal to the 95th percentile for their age group, the term "overweight" is used. BMI-for-age and weight-for-size charts may utilize the 85th percentile to identify those who are at danger of becoming overweight.

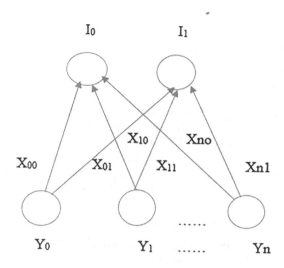

FIGURE 7.2
SOM.

Adults are classified differently than children and adolescents when it comes to being overweight or underweight. Adults are classified as overweight or underweight based on set BMI cut points obtained from morbidity and mortality data. Those with a BMI of 18.5 or more but less than 30.0 had a greater relative mortality risk than adults with a BMI of 18.5 or more but less than 30.0. People of all ages and genders may use the BMI calculator to see whether they are at risk of obesity (Figure 7.2).

a. SOM Process

One of the most often used neural models is the SOM. It's coming to the competitive learning system area. The SOM is based on unsupervised learning, which means that no human intervention is necessary during training and that only a little quantity of input data information is required. The SOM could be used, for example, to cluster input data membership. The SOM is also known as the Self-Origination Feature Map since it may be used to find features that are intrinsic to the situation. The architecture of every sample's SOM with two clusters as well as n input characteristics is shown in Figure 7.2.

The following are the phases of the SOM technique that utilizes the below algorithm.
Algorithm

Step 1: Initialization

Initialize the weight vectors w_i with random values.

Step 2: Sampling

Make a practice training input vector x from the input space.

Step 3: Matching

For simplicity, let us assume that $I(x)$ win since it has the lowest weight vector and is located closest to the input vector in terms of distance.

$$F_i(x) = \sum_{j=1}^{D} (x_j - w_{ij})^2 \tag{7.2}$$

Step 4: Updating

    Make use of the weight-loss method.

Step 5: Continuation

    Repeat steps 2 and 3 until the feature map is no longer changing.

1. **Initialize Random Weight Vector:** Random weight vector is created in this stage. For iterative approaches such as ANN, obtaining a decent initial approximation is a well-known difficulty. Although random weight initialization is common, principle component initialization, which takes advantage of the data's first principal component space, has gained popularity due to the results' repeatability. The input space should be the same size as the weight vector dimension/size of neurons.

2. **Pick a Random Input:** Use the input that was chosen at random.

3. **Compute Winner Neuron Using Euclidean Distance:** Neuron is assigned a Euclidean distance based on the random selection of an input. The computation of the two neurons is shown in the table below. The winning neuron is the one with the smallest distance. W1 is the winning neuron in this case. The Best Matching Unit (BMU) is the winning neuron. Other distance metrics can be used to compute BMI instead of Euclidean distance. The size of the BMU's neighborhood is calculated in this stage, with the size based on an exponential decay function. The neighborhood surrounding the BMU is shrinking over time.

4. **Update Neuron Weights:** This stage modifies both the BMU and the weights of the neighborhoods in order to investigate the most similar neuron to the input vector and to anticipate a big change for neighbors closer to the BMU. The farther the distance from the BMU, the less the neighbors learn (insignificant change in the weights).

5. Iterate Steps 1–4 until the locations of the neurons are stabilized, and then repeat Steps 2–4 until the training is complete.

Despite being a widely used and easily interpretable unsupervised technique for demonstrating the similarity between data inside a low-dimensional grid, SOM has some limitations: Insufficient or unnecessary data can cause clusters to become more random; hence, the SOM requires a large quantity of data to generate meaningful clusters.

With categorical data, SOM has difficulty grouping. After translating the data into an indicator variable, the SOM is usually applied to the categorical variable (like a binary variable). Similarity qualities across categorical features may lose crucial information as a result of the transformation, and a trained SOM may not be able to offer the correct topological aspects of the data.

The purpose of the SOM is to store a large number of input vectors x by identifying a portable number of cases that closely approximate the original input space. Vector quantization theory, which is inspired by dimensionality reduction or data compression, provides theoretical underpinning for this notion. The precision of the approximation is determined by the total squared distance in real impact (Figure 7.3).

$$D = \sum x \left\| y - w_{I(x)} \right\|^2 \tag{7.3}$$

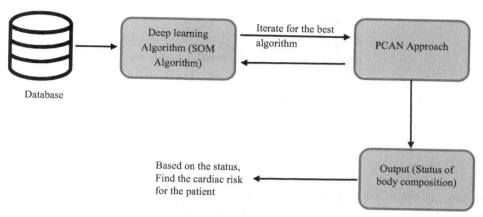

FIGURE 7.3
Processing.

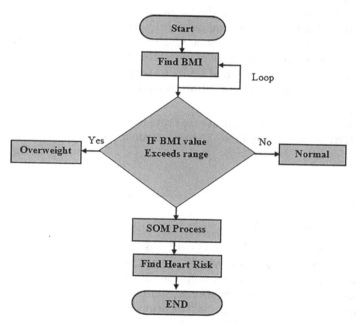

FIGURE 7.4
Flowchart of SOM process.

The diagram above depicts how data are extracted from a dataset and how they may be used in a deep learning algorithm, as well as how the iteration process can be used to select the optimal algorithm. After deciding on the optimum methodology for determining people's cardiac risk using the BMI app, the SOM algorithm was used to determine cardiac risk in this study. The BMI application is mostly used to determine heart risk. It uses data from a dataset to determine a person's weight and height and then uses that information to calculate heart risk (Figure 7.4).

TABLE 7.3

BMI Category

BMI Category	Female	Male
Extremely weak	80	70
Weak	10	15
Normal	45	40
Overweight	35	30
Obesity	70	55
Extreme obesity	90	115

We can get the SOM weight update algorithm by using gradient descent-style mathematics, which verifies that it generates a good approximation to the input data.

a. Validation

Validation diagnosis must be backed up by clinical criteria that are widely accepted in the medical community. Clinical criteria that are widely recognized are usually based on authoritative professional guidelines, consensus, or evidence. If a person can look at the indications, symptoms, findings, and documentation, then identify a gap and apply the relevant clinical indicators to construct a query for a precise diagnosis, then the converse must be true – a person must be able to notice the information of a diagnosis or condition as well as isolate.

b. Evaluation

The accuracy, recall, precision, and error rate of the SOM technique are calculated using the BMI presented in Table 7.3 to evaluate its performance results.

7.4 Results and Discussion

Each of the 700 entries in the BMI dataset contains the following four characteristics: The first four considerations are gender, weight, height, and BMI. Using the index, it is possible to evaluate whether a person is exceptionally thin or highly fat (5). The data were utilized to create a multiclass classification system that categorized individuals based on their physical makeup. For multiclass classification, we used the "SOM" method. "Risk" was added to the multiclass "BMI" dataset to reduce it to a binary classification problem using a calibrated classification approach, with the logic that "Risk" = 0 if (very weak, weak, normal weight) and 1 if (very strong, strong, strong). We included records with ages between 20 and 60 in the preprocessing of the data and omitted records containing the feature "children."

BMI Category

A comparison of BMI and index column values yields the following findings:

"0 = Extremely weak: BMI < 16

1 = Weak: 16 < BMI < 18.5

2 = Normal: 18.5 < BMI < 24.9

3 = Overweight: 25 < BMI < 29.9

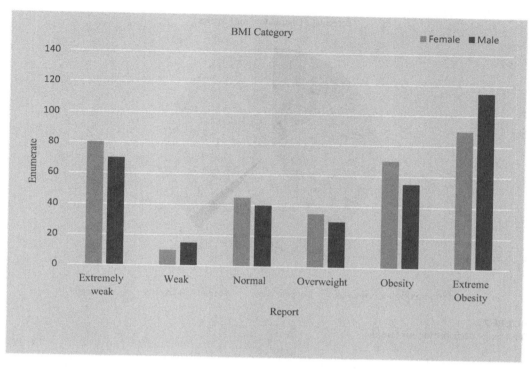

FIGURE 7.5
BMI category for male and female.

4=Obesity: 30<BMI 34.9

5=Obesity with a BMI of greater than 35."

A BMI of less than 18.5 indicates that an individual is underweight and may require weight increase. They should get advice from a physician or a dietician. When it comes to weight, a BMI of 18.5–24.9 is considered normal for someone of a certain height. Keeping a healthy weight might help them avoid serious health issues. An overweight individual has a BMI of 25–29.9. If you have health problems, your doctor may advise you to lose weight. They should seek the opinion of a doctor or nutritionist. Having a BMI of 30 or more is considered obese. Their health may be jeopardized if they do not lose weight. They should get advice from a physician or a dietician (Figure 7.5).

Men's and women's BMI categories are shown in the table above. Women and men in this BMI range may be characterized as either extremely weak, weak, normal weight, overweight, obese, or very obese. As previously stated in this chapter, this category has three values: obesity, overweight, and normal. The graph above depicts the report and enumeration of the BMI category for both men and women. The BMI is computed using the person's height, weight, and age (Figure 7.6 and Table 7.4).

Body fat levels in women are often higher than in men. The amount of fat in your body fluctuates as you get older. The following are the ideal body fat percentages for each age group:

At the age category of 20–29, the body fat level under 14% means the body health is extremely weak, the body fat level 14.2%–16.4% means the health condition is weak, the body

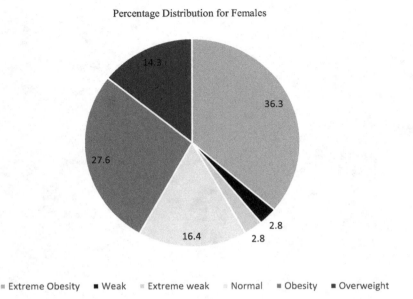

FIGURE 7.6
Percentage distribution for female.

TABLE 7.4

Body Fat Percentage for Female

BMI Category	Percentage
Extreme obesity	36.3
Weak	2.8
Extremely weak	2.8
Normal	16.4
Obesity	27.6
Overweight	14.3

fat level 16.7%–19.2% means the health condition is normal, the body fat level 19.4%–22.6% means the health condition is overweight, the body fat level 22.7%–27% means the health condition is obesity, and the body fat level 31% means the health condition is extreme obesity.

At the age category of 30–39, the body fat level under 14% means the body health is extremely weak, the body fat level 14%–17.2% means the health condition is weak, the body fat level 17.3%–20.7% means the health condition is normal, the body fat level 20.8%–24.5% means the health condition is overweight, the body fat level 24.6%–29.3% means the health condition is obesity, the body fat level 33% means the health condition is extreme obesity.

At the age category of 40–49, the body fat level under 14% means the body health is extremely weak, the body fat level 14%–19.6% means the health condition is weak, the body fat level 19.9%–23.9% means the health condition is normal, the body fat level 24%–27.7% means the health condition is overweight, the body fat level 27.8%–32% means the health condition is obesity, and the body fat level 32.6% means the health condition is extreme obesity.

At the age category of 50–59, the body fat level under 14% means the body health is extremely weak, the body fat level 14%–22.6% means the health condition is weak, the body

Percentage Distribution for Males

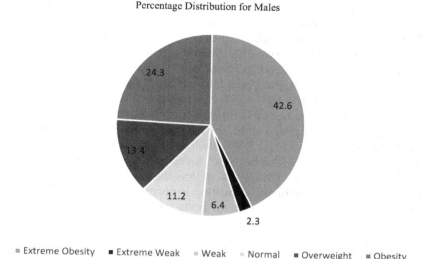

FIGURE 7.7
Percentage distribution for male.

TABLE 7.5

Body Fat Percentage for Male

BMI Category	Percentage
Extreme obesity	42.6
Extremely weak	2.3
Weak	6.4
Normal	11.2
Overweight	13.4
Obesity	24.3

fat level 22.7%–27.3% means the health condition is normal, the body fat level 27.5%–30.5% means the health condition is overweight, the body fat level 30.6%–34.7% means the health condition is obesity, and the body fat level 34.8% means the health level is extreme obesity.

At the age category of over 60, the body fat level under 14% means the body health is extremely weak, the body fat level 14%–23.4% means the health condition is weak, the body fat level 23.5%–30% means the health condition is normal, the body fat level 30.2%–31.5% means the health condition is overweight, the body fat level 31.5%–35.7% means the health condition is obesity, the body fat level 35.8% means the health condition is extreme obesity (Figure 7.7 and Table 7.5).

Men should maintain a healthy body fat percentage as well. Men have slightly lower optimal fat percentages than women. The following are the ideal body fat percentages for each age group:

At the age category of 20–29, the body fat level under 8% means the body health is extremely weak, the body fat level 8%–10.6% means the health condition is weak, the body fat level 10.7%–14.9% means the health condition is normal, the body fat level 15%–18.7% means the health condition is overweight, the body fat level 18.8%–23.2% means the health condition is obesity, the body fat level 23.5% means the health condition is extreme obesity.

At the age category of 30–39, the body fat level under 8% means the body health is extremely weak, the body fat level 8%–14.6% means the health condition is weak, the body fat level 14.7%–18.3% means the health condition is normal, the body fat level 18.4%–21.4% means the health condition is overweight, the body fat level 21.5%–25% means the health condition is obesity, the body fat level 25.3% means the health condition is extreme obesity.

At the age category of 40–49, the body fat level under 8% means the body health is extremely weak, the body fat level 8%–17.5% means the health condition is weak, the body fat level 17.6%–20.7% means the health condition is normal, the body fat level 20.8%–23.5% means the health condition is overweight, the body fat level 23.6%–27% means the health condition is obesity, the body fat level 27.2% means the health condition is extreme obesity.

At the age category of 50–59, the body fat level under 8% means the body health is extremely weak, the body fat level 8%–19.5% means the health condition is weak, the body fat level 19.7%–22.4% means the health condition is normal, the body fat level 22.5%–24.7% means the health condition is overweight, the body fat level 24.9%–27.9% means the health condition is obesity, the body fat level 28.4% means the health condition is extreme obesity.

At the age category of over 60, the body fat level under 8% means the body health is extremely weak, the body fat level 8%–19.9% means the health condition is weak, the body fat level 20%–22.8% means the health condition is normal, the body fat level 23%–25.7% means the health condition is overweight, the body fat level 25.9%–28.6% means the health condition is obesity, the body fat level 28.8% means the health condition is extreme obesity.

We can see that males are more amazingly hefty than females, while females are stouter and more typical, and females are more overweight than males in the dispersion of individuals by status as per sexual orientation (Figure 7.8 and Table 7.6).

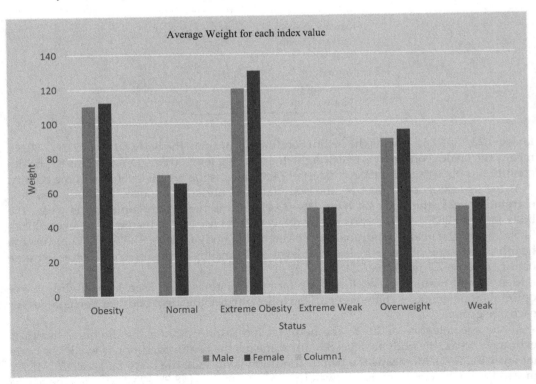

FIGURE 7.8
Average weight for both male and female.

TABLE 7.6

Distribution of People by Weight

BMI Category	Male	Female
Obesity	110	112
Normal	70	65
Extreme obesity	120	130
Extremely weak	50	50
Overweight	90	95
Weak	50	55

TABLE 7.7

Distribution of People by Height

BMI Category	Male	Female
Obesity	160	165
Extreme obesity	150	155
Weak	180	178
Extremely weak	185	182
Overweight	172	175
Normal	175	170

In the above table, this research explains weight distribution of male and female gender. Based on the weight of both male and female, their body fat level and the heart risks are identified. If the person's weight is beyond the age level, it means the person is highly likely to be affected by heart risks. The person will control the idea that is given by the doctors; otherwise, it will cause death (Table 7.7).

In the above table, this research explains the height distribution of male and female gender. Based on the height of both male and female, their body fat level and the heart risks are identified. If the person's height is beyond the age level, it means the person is highly likely to be affected by heart risks. The person will control the idea that is given by the doctors; otherwise, it will cause death.

Based on the height and weight, only the BMI value is calculated. If the height and weight range is beyond the age of the person, irrespective of gender, he/she is highly likely to be affected by the cardiovascular risks. In this research, it is mainly identified that compared to females, males are mainly affected by fats they have.

Figure 7.9 shows the results of the dataset's regression analysis, which reveals a positive relationship. In this study, SOM algorithms were utilized to divide the records into two categories: obese (0) and non-obese. Obesity and diabetes are linked in both males and females, according to this study. If your BMI is within the normal range, you'll be able to prevent some of the world's most common cardiac problems. This document provides 85% correct information. The blood glucose and blood pressure results are shown in the graphs below (Table 7.8).

Figure 7.10a shows that many biometric examinations also monitor your blood glucose (commonly known as "blood sugar"). This is a crucial measure of your body's ability to utilize the carbohydrates you ingest. High blood glucose levels can suggest prediabetes or diabetes, depending on the level. Testing your blood glucose levels in the present or during the previous two or three months with a hemoglobin A1c test is one way to determine your typical blood glucose level. Depending on the test, the normal and high ranges might vary greatly. As long as you do not have type 1 diabetes, in which your body does not create

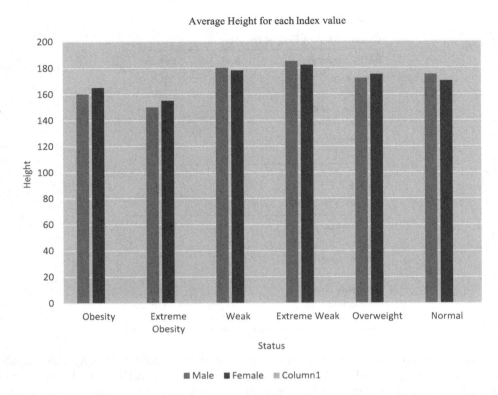

FIGURE 7.9
Average height for both male and female.

TABLE 7.8

Evaluation of SOM

	Gender	Height	Weight	Index	Status
0	Male	174	96	4	Obesity
1	Male	189	87	2	Normal
2	Female	185	110	4	Obesity
3	Female	195	104	3	Overweight
4	Male	149	61	3	Overweight
5	Male	189	104	3	Overweight
6	Male	147	92	5	Extreme obesity
7	Male	154	111	5	Overweight
8	Male	174	90	3″	Overweight

enough insulin to break down sugar, your doctor will recommend a plan to decrease your blood sugar levels and prevent or cure type 2 diabetes. The inability of your body to properly use insulin causes type 2 diabetes, which results in an excess of glucose in your body.

Figure 10b indicates that blood pressure is extremely high and that medication is needed to bring it under control. If you're on the verge of becoming obese, lifestyle and dietary modifications may be enough to bring your statistics down to a healthy level. Among the modifications are the following:

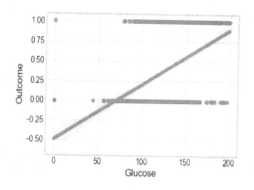

(a) The relationship between the result (obesity) and blood glucose levels

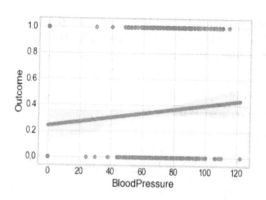

(b)The link between the result (obesity) and blood pressure

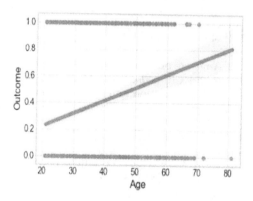

(c) Relationship between age and the result (obesity)

FIGURE 7.10
(a–c) Evaluation of obesity.

- Losing weight in order to maintain a healthy BMI.
- Eating more fresh fruits and switching to a more plant-based diet in place of junk and fried foods.
- Your daily salt consumption should not exceed 2,300 mg/day. (Less than 1,500 mg is even better.)
- Improving your ability to cope with stress may help lower your blood pressure.

7.5 Conclusions

In this chapter, a SOM-based approach was proposed for detecting cardiac risk in ECG signals. The performance of the heart rate may be evaluated in our method using the height, weight, and age of the individual datasets. The experimental results suggest that the BMI may accurately predict blood glucose, blood pressure, and heart risk depending on the BMI number. In comparison with the existing PCAN approach, our technique accurately predicted heart risk.

References

[1] Jindal, K., Niyati, B., Rana, P.S. Obesity prediction using ensemble machine learning approaches. In Sa, P.K., Bakshi, S., Hatzilygeroudis, I.K., Sahoo, M.N. (eds.) *Recent Findings in Intelligent Computing Techniques* (pp. 355–362). Springer Nature, Singapore, 2018.

[2] Pantanowitz, A., E. Cohen, P. Gradidge, N.J. Crowther, V. Aharonson, B. Rosman, D.M. Rubin. Estimation of body mass index from photographs using deep convolutional neural networks. *Informatics in Medicine Unlocked*, 100727, 1–15, 2021.

[3] Dong, D., et al. The role of imaging in the detection and management of COVID-19: A review. *IEEE Reviews in Biomedical Engineering*, 14, 16–29, 2021.

[4] Ramasamy, M.D., Periasamy, K., Krishnasamy, L., Dhanaraj, R.K. Multi-disease classification model using Strassen's half of threshold (SHoT) training algorithm in healthcare sector. *IEEE Access*, 9, 112624–112636, 2021.

[5] Han, H., Linlin, X., Rui, L. Analysis of heart injury laboratory parameters in 273 COVID-19 patients in one hospital in Wuhan, China. *Journal of Medical Virology*, 92, 819–823, 2020.

[6] Marcel, R., Sobh, J.F., Saul, P. Beat detection and classification of ECG using self-organizing maps. In *Proceedings of the 19th Annual International Conference of the IEEE Engineering in Medicine and Biology Society*, 1, 89–91, 1997.

[7] Shufelt, C.L., A. Kim, S. Joung, L. Barsky, C. Arnold, S. Cheng, S. Dhawan, et al. Biometric and psychometric remote monitoring and cardiovascular risk biomarkers in ischemic heart disease. *Journal of the American Heart Association*, 9(18), e016023, 2020.

[8] Odinaka, I., P.-H. Lai, A.D. Kaplan, J.A. O'Sullivan, E.J. Sirevaag, S.D. Kristjansson, A.K. Sheffield, J.W. Rohrbaugh. ECG biometrics: A robust short-time frequency analysis. In *2010 IEEE International Workshop on Information Forensics and Security*, pp. 1–6. IEEE, 2010.

[9] S. Jayachitra, A. Prasanth, Multi-feature analysis for automated brain stroke classification using weighted Gaussian Naïve Baye's classifier. *Journal of Circuits, Systems, and Computers*, 30, 2150178:1–26, 2021.

[10] Pirbhulal, S., H. Zhang, S.C. Mukhopadhyay, C. Li, Y. Wang, G. Li, W. Wu, Y.-T. Zhang. An efficient biometric-based algorithm using heart rate variability for securing body sensor networks. *Sensors*, 15(7), 15067–15089, 2015.

[11] Okorodudu, D.O., M.F. Jumean, V.M. Montori, A. Romero-Corral, V.K. Somers, P.J. Erwin, F. Lopez-Jimenez. Diagnostic performance of body mass index to identify obesity as defined by body adiposity: A systematic review and meta-analysis. *International Journal of Obesity*, 34(5), 791–799, 2010.

[12] Flint, A.J., K.M. Rexrode, F.B. Hu, R.J. Glynn, H. Caspard, J.E. Manson, W.C. Willett, E.B. Rimm. Body mass index, waist circumference, and risk of coronary heart disease: A prospective study among men and women. *Obesity Research & Clinical Practice*, 4(3), e171–e181, 2010.

[13] Akhter, N., Tharewal, S., Kale, V., Bhalerao, A., Kale, K.V. Heart-based biometrics and possible use of heart rate variability in biometric recognition systems. In Chaki, R., Cortesi, A., Saeed, K., Chaki, N. (eds.) *Advanced Computing and Systems for Security* (pp. 15–29). Springer, New Delhi, 2016.

[14] Wang, D., Si, Y., Yang, W., Zhang, G., Liu, T. A novel heart rate robust method for short-term electrocardiogram biometric identification. *Applied Sciences*, 9(1), 201, 2019.

[15] Castelli, W.P., R.D. Abbott, P.M. McNamara. Summary estimates of cholesterol used to predict coronary heart disease. *Circulation*, 67(4), 730–734, 1983.

[16] Shahid, S.U., Sarwar, S., 2020. The abnormal lipid profile in obesity and coronary heart disease (CHD) in Pakistani subjects. *Lipids in Health and Disease*, 19, 1–7.

[17] Wang, K, Yang, G, Huang, Y, Yin, Y. Multi-scale differential feature for ECG biometrics with collective matrix factorization. *Pattern Recognition*, 1, 102:107211, 2020.

[18] Huang, Y., Yang, G., Wang, K., Liu, H., Yin, Y. Learning joint and specific patterns: A unified sparse representation for off-the-person ECG biometric recognition. *IEEE Transactions on Information Forensics and Security*, 16, 147–160, 2020.

[19] El Sheikh, H., Shalaan, A., Abdelazeem, E. Fetal echocardiography as a screening method for detection of congenital heart disease. *Benha Medical Journal*, 38, 11–21, 2021.

[20] Rizzo, P., Dalla Sega, F.V., Fortini, F., Marracino, L., Rapezzi, C., Ferrari, R. COVID-19 in the heart and the lungs: Could we "Notch" the inflammatory storm? *Basic Research in Cardiology*, 115(3), 31, 2020.

[21] Widiyaningtyas, T., Zaeni, I.A.E., Wahyuningrum, P.Y. Self-organizing map (SOM) for diagnosis coronary heart disease. In *2019 4th International Conference on Information Technology, Information Systems and Electrical Engineering (ICITISEE)* (pp. 286–289). IEEE.

[22] Chatterjee, A., Gerdes, M.W., Martinez, S.G. Identification of risk factors associated with obesity and overweight—A machine learning overview. *Sensors*, 20(9), 2734, 2020.

[23] Steven, A., J.M. Irvine. Heartbeat biometrics: A sensing system perspective. *International Journal of Cognitive Biometrics*, 1(1), 39–65, 2012.

[24] Melin, P., J.C. Monica, D. Sanchez, O. Castillo. Analysis of spatial spread relationships of coronavirus (COVID-19) pandemic in the world using self-organizing maps. *Chaos, Solitons& Fractals*, 138, 109917, 2020.

[25] Drozek, D., A. DeFabio, R. Amstadt, G.Y. Dogbey. Body mass index change as a predictor of biometric changes following an intensive lifestyle modification program. *Advances in Preventive Medicine*, 2019, 2019.

[26] Vakli, P., R.J. Deák-Meszlényi, T. Auer, Z. Vidnyánszky. Predicting body mass index from structural MRI brain images using a deep convolutional neural network. *Frontiers in Neuroinformatics*, 14, 10, 2020.

[27] Mason, J., R. Dave, P. Chatterjee, I. Graham-Allen, A. Esterline, K. Roy. An investigation of biometric authentication in the healthcare environment. *Array*, 8, 100042, 2020.

8

Embedded Medical IoT Devices for Monitoring and Diagnosing Patient Health in Rural Areas Peoples Using IoMT Technology

V. Karuppuchamy
Kongunadu College of Engineering and Technology

S. Palanivel Rajan
M. Kumarasamy College of Engineering

C. Manikandan
Pandian Saraswathi Yadav Engineering College

CONTENTS

DOI: 10.1201/9781003256243-8

8.1 Introduction

Nowadays, the healthcare industry is under rapid growth and provides employment and revenue [1]. Some years before, patient health abnormalities have been found out by the doctor in the hospital itself. The patient needs to stay throughout the time in the hospital to cure the disease. The fact of the result is healthcare costs have increased. The technology enhancement turns the advancement through mobile phone monitors to some of the health systems. Our watch regularly monitors the heartbeat, temperature, blood pressure, and several clinical analyses [2,3]. The data can be transferred using wireless technologies.

The IoT technology does not only give independence, but also helps humans interact with the external world. We have used many useful protocols and algorithms to communicate in an efficient way through the Internet [4]. The IoT technology can also be used in the agricultural field [5–7], smart grid [8], home, and healthcare [9]. The improvement of IoT technologies is it gives the maximum accuracy at a lower cost and it predicts the features in a good way. With the advancement of mobile and computer technology, we can easily use wireless communication technology. Since the economy has increased and the rapid IoT revolution [10].

Different communication protocols are available nowadays, such as Wi-Fi, Bluetooth, and IEEE 802.11. A variety of sensors are used to collect the information from the human body and are also very small and wearable [11]. The data transmission between the models is maintained in a more secure way. Most of the research reported on healthcare monitoring and security [12,13]. In recent trend, more employability might provide new technologies and policies. The IoT framework applied to the healthcare industry is integrated with cloud computing technology. System or network highly coupled with the healthcare device, publisher, broker, and subscriber [14]. The publisher wears the device and continuously sends the information to the broker environment. If the distance is very less, Bluetooth/Wi-Fi technology is used.

The second connectivity is GPRS; in this, distance is not limited. The subscriber can read the data and know the current information about the patient. The subscribers are medical researchers, pharmaceutical scientists, and insurance contributors. The system must accept the changes and adopt past HIoT systems. It is very tough to manipulate a new healthcare environment, and the device must satisfy the medical research rules and government policies, and IEEE standards, network topologies pursue system and convention in conclusion and procedure [15–17] (Figure 8.1).

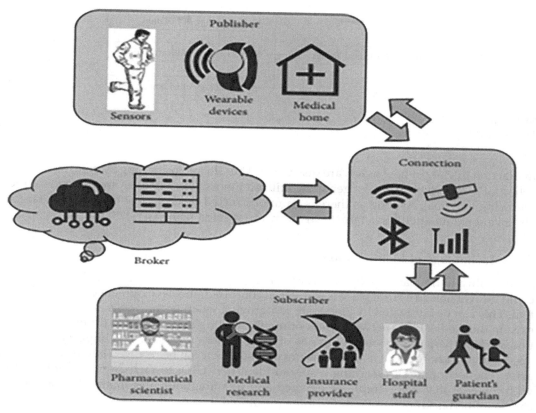

FIGURE 8.1
Structure of HIoT framework.

For IoMT infrastructure, technology design is most important. This is because using certain technologies can improve an IoT system's capability. Numerous cutting-edge technologies are being used to incorporate different medical care applications with an IoT framework. ID innovation, correspondence innovation, and area innovation are without doubt the three essential classifications where these advances fall.

8.2 HIoT Technology

It is very tough to develop a HIoT system. The system has to improve the performance system model [18,19]. Since we associate the variety of protocols and communication models being accepted, this technology is classified into three types: technology of location, technology of identification, and technology of communication (Figure 8.2).

8.2.1 Location Technology

The ongoing area frameworks are consistently utilized in the medical care framework to screen and find the whereabouts of an article. It additionally keeps a record of the treatment

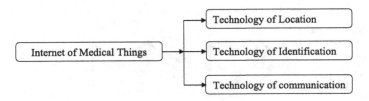

FIGURE 8.2
Taxonomy of IoT technology.

interaction in view of how assets are dissipated. The Global Positioning System, otherwise called GPS, is perhaps the most generally utilized innovation. Satellites are utilized for the following end goal: As long as there is an unobstructed field of view between the target and five separate satellites, a target can always be spotted using GPS [23–26].

8.2.2 Technology of Identification

Accessibility to the patient's health-related information from the designated devices that could be available in a rural place is a practical concern in the development of an IoMT system. This can be cultivated by obviously recognizing whether the nodes and sensors that exist in the medical care network are continually sending information to different applications. When there is a problem with one of these nodes or sensors, it is important for the application to know about it as soon as possible so that it can be resolved before any issues arise. The technique of assigning a special identifier (UID) to each approved substance with the goal that it is handily recognized and unambiguous information transmission can be refined is known as distinguishing proof [20,21].

8.2.3 Technology of Communication

Short-range protocol communication technology is for establishing connections between objects within a limited range, whereas long-range communication technology of communication typically supports communication over a vast space, comparable to correspondence in the middle of the base station and the focal hub of BAN. Correspondence conventions include RFID and Bluetooth. Bluetooth frequency is 2.4 GHz, and its range is 100 m. Zigbee operates in the frequency range of 2.4 GHz, adopts the mesh network topology, consumes less power, and has a high transmission rate. NFC works on the basis of electromagnetic induction between two antennas. NFC works in two modes: active mode and passive mode. In the active mode, one device produces the induction, whereas in the passive mode, two devices produce the induction. The advantage is wireless communication and is used for a very short range of communication [22].

8.3 Applications and Services of IoMT

Medical gadgets can now do meaningful evaluations that doctors could not even accomplish just a few decades ago. It is all because of the recent advancements in IoT technology. This has also allowed healthcare organizations to communicate to as many individuals

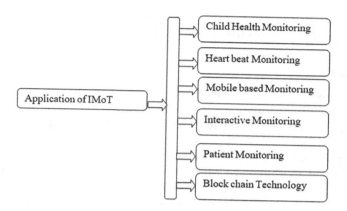

FIGURE 8.3
IMoT services.

simultaneously providing benefits to the public. The exploit of cloud computing has tremendously enhanced the performance as well as information exchange between patients and doctors. As an outcome, the patient's involvement in the primary treatment has been improved, whereas the patient's financial strain has also reduced (Figure 8.3).

The present significant influence of IoT is driving the creation of IoMT applications, which include disease diagnostics, individual concern for geriatric people, fitness, health management, and diabetes complications management. It is classified into two distinct fundamental categories: services and applications. The former reflect the values that are implemented in the creation of an IoMT device, while the latter refer to healthcare applications that are used in either diagnosing a specific medical condition, or measuring health parameters. The services and uses of IoMT are described in detail in the following sections.

8.3.1 Services

By giving solutions to numerous healthcare concerns, solutions and theories have altered the healthcare business. With growing healthcare requirements and technological advancements, more services are being introduced on a daily basis. These are now becoming an important part of the IoMT system design process. In an IoMT environment, each business offers a set of healthcare solutions. These concepts/services aren't really defined in a unique way. The applications are what make IoMT systems stand out. As a result, defining each notion in a broad sense is difficult. However, the following section describes a few of the main extensively used in the medical field.

8.3.1.1 Ambient Assisted Living

The subset of artificial intelligence is ambient assisted living, which is related to the IoT [27]. By using this technology, older adults live independently in a home with a safe environment. Ambient assisted living provides a way for the observance and dominance of these patients in real time and makes support in the event of a medical emergency. AI technologies, huge amounts of data investigation, deep learning, also their utilization in the medical services industry. Researchers are looking into three necessary domains of AAL: recognition of activity, monitoring, and recognition of environment. Nevertheless, activity recognition drew the most understanding because it manages to perceive possible risks or critical well-being circumstances that could harm senior patients well indeed [28–37].

8.3.1.2 Mobile IoT

Integration of mobile computing and cloud computing, sensors are used to trach the information of patient health information, track patient fitness data and different physiological answers is known as cell IoT. A sturdy hyperlink among non-public place networks and cellular networks is established so that you can transport a dependable primarily Internet-based healthcare service. 4G and 5G provide edgeless communication to transform the data from one place to another place. More research reported on mobile computing healthcare applications [38–41]. When an abnormal heartbeat is detected, an alert is sent to the patient and nearby doctors [42]. Security concern is the most important thing in the IoT system, which needs focus on signal lagging and error rate. Power failure, connectivity problem, and transmission problem are the drawbacks [43].

8.3.1.3 Wearable Devices

The wearable generation assists it less complicated for healthcare providers and patients to deal with a wide form of illnesses at a decreased price. These unobtrusive gadgets can be made through blending numerous sensors with human-wearable add-ons together with watches, wristbands, necklaces, shirts, footwear, and purses so far. The linked nodes collect the patient's fitness data and so on. After that, the information is published to the server/databases. Through health programs, several wearable gadgets also are linked to mobile telephones. The usage of these kinds of wearable gadgets and mobile computing in real-time monitoring changed into pronounced in pretty some studies (Figures 8.4–8.6).

8.3.1.4 Cognitive Computing

Mental figuring alludes to the means of perusing an aggravation similar to the human psyche. With the current upgrades in sensors and manufacturing knowledge, sensors are productively incorporated into the system. Mental processing in an IoT machine makes it more straightforward to peruse stowed away styles that can be found in a colossal measure. Moreover, incorporates the limit of different sensors to get well-being generally adjusted and its current circumstance. In a mental IoT organization, all sensors work together with various brilliant gadgets and give effective wellness administrations.

Involving mental processing in an IoT machine empowers medical care experts to successfully remark on persistent data and deal with the right treatment. In completely smart well-being observing machine-dependent principally upon EEG was proposed which utilizes mental figuring to decide the patient's neurotic conditions+EEG data, alongside different sensor data just as discourse, motions, outline developments, and looks, were utilized to evaluate a patient's circumstance. It furthermore allows emergency help if there should arise an occurrence of pathogenic conditions.

8.3.1.5 Reaction of Drug

Drug reaction could be depicted as a point effect of attractive medication. The reaction can endure increasing each after alone piece and after a long stretch of association. This, in a like manner, can be achieved due to the point sway while specific holders. Now not

FIGURE 8.4
HIoT services.

depend on the medicine type or contamination and changes starting with one individual and then onto the next other. In an IoT-based ADR device, a very precise identifier/scanner tag is used to pick out every drug at the affected individual terminal. Drug compatibility with the affected individual body can be set up by the utilization of reasonable medication measurements instruments. The measurements hardware shows the hypersensitive profile of each impacted individual with the use of e-health documents. In a comparable report, an IoT remedy-based absolutely certainly (prescADE) drug-unfriendly occasion apparatus has been proposed, which additionally can embellish impacted individual well-being through diminishing ADE.

8.3.1.6 Blockchain

The exchange of data between medical equipment and explicit care partner companies plays a very important role. One of the main problems with exchanging data securely is fact fragmentation. The blockchain age is also fueling collaboration between health

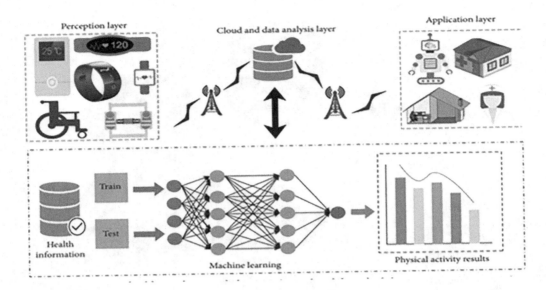

FIGURE 8.5
MI-based healthcare system.

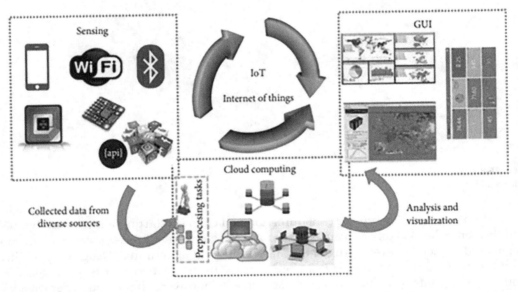

FIGURE 8.6
IoT environment.

professionals and companies doing qualitative analysis. Secure and elegant transmission occurs within the blockchain technology. The first one is an immutable "ledger," over which humans gain access and control. Second, blockchain could be a distributed technology and operates at the same time with n number of devices, and so forth agreement, information swap policy with vice versa.

The developer has the right to access the entire medical report and store it in blockchain with cloud technology. The doctors are allowed to access the report if the manager permits.

In current years, many blockchain technologies have been developed to aid in the business and other medical clinical pathways. Dialect et al. created a mobile application, in which the nurse can provide the authority to share their patient information. Without violating the rules and regulations, the patient can share the details with others for consulting.

8.3.1.7 Child Health Information

Well-being information of a child can be built that arranges with bringing issues to light of the prosperity of a kid. The primary objective of CHI is to instruct and enable youngsters and their older folks on the general soundness of the kid just as on the qualities of natural interaction, enthusiastic and mental state, and conduct. The IoT application has assisted analysts in accomplishing this with the occasion of a stage that will screen and direct a youngster's well-being. Nigar and Chowdhury fostered an IoT-based system in which a youngster's psychological and actual state is observed.

What's more, fundamental measures are taken with the assistance of specialists just in case of a nursing crisis. In an incredibly comparable review, an IoT-based clinical organization was fostered that interfaces a clinical gadget to a portable application. The framework gathers five totally unique body boundaries: tallness, temperature, SpO_2, weight, and pulse. These data are shaped open to specialists and well-being experts from the application. The utilization of an m-well-being administration was to notice the dietary patterns of youngsters by instructors and guardians. The application has been utilized to acquire the upsides of touchy natural cycles in youngsters (Figure 8.7).

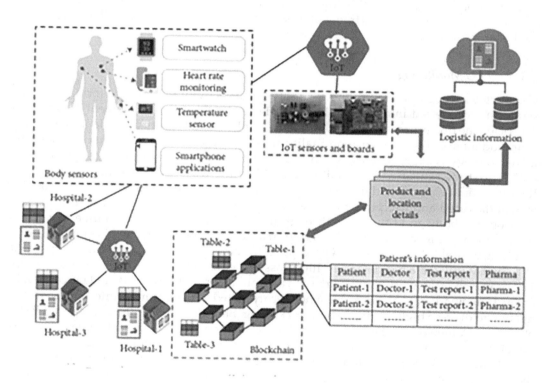

FIGURE 8.7
Health monitoring system based on blockchain technology.

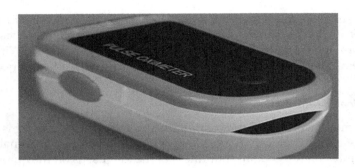

FIGURE 8.8
IoT oxygen meter.

8.3.2 Applications

Administrations/ideas have utilized the improvement of completely free IoT programs. Analysts in these fields have extended different thoughts onto the vector of humankind. So, the smartest thoughts are designer-driven, while programs are client-driven simultaneously. The fast improvement of the IoT time has prompted the improvement of less expensive and more straightforward wearable sensors, wearables, and clinical gadgets. These offices are additionally used to accumulate data about the impacted individual, analyze infections, reveal the well-being status of impacted people, and create clinical crisis cautions (Figure 8.8). The assortment of new units accessible in the market has been examined later in the area.

8.3.2.1 ECG Monitoring

The electrocardiogram is used to measure the heart activity in a picture format, and the signals are converted into a graph through electrodes. It reads the moment of atria and ventricle. The doctors measure the rhythm of your heart and identify the blood flow of the heart and muscle. By using an abnormality, the doctor predicts cardiac abnormalities. The IoT technology is expressively used to find the abnormalities in the early stages through ECG. The study explained IoT monitoring systems and wireless data acquisition systems. Regular monitoring method that detects the abnormality in real time. A low-power device is manufactured and fixed into the jacket. A lot of small sensors are used to read the signal from the body and transmit it through the GPRS or Bluetooth. The numeric value is processed by the machine learning algorithm and monitored by the mobile application.

A little wearable low-power EKG perception framework was designed and incorporated in a shirt. It utilized a bio-potential chip to assemble reasonable quality ECG information. The proposed framework could work with a minimal force of 5.2 mW. Period perception in an IoT framework might be possible when incorporated with enormous information investigation to oversee higher information stockpiling. Bansals and Gandhi arranged an EKG-observing framework that might deal with long and non-stop following through methods for joining the idea of nanoelectronics. More value to know in creators endeavored to clear up solidarity related cardiogram looking gadgets. A powerful method is used to detect the streamline of the signal. This gadget became intended to give the term following to matured victims through the method of routinely looking at their cardiogram and estimating device data.

8.3.2.2 Monitoring the Glucose Level

Diabetes is the maximum sugar or glucose occupied maximum in the body and imme-diately increases unexpected illness in individuals. Overall, there are three fundamental sorts of hereditary issues, specifically the hereditary sickness type I, the pair of polygenic infections type An, and the physiological state of diabetes. The sickness and its sorts are additionally perceived after three tests, specifically the arbitrary glucose test, the fast glu-cose test, and the oral aldohexose resilience test. Smear "trailed by a glucose test. The new advancement of IoT advances has been utilized in the improvement of a wide assort-ment of handheld blood glucose meters that are painless, helpful, advantageous, and safe. A harmless m-IoT-based blood glucose meter was produced for the fundamental quantita-tive investigation of the noticed blood glucose esteems. Handheld sensors and carers are incorporated into the IPv4 network.

8.3.2.3 Temperature Monitoring

Holding homeostasis and is a basic piece of numerous analytic cycles is human tempera-ture, also an additional an in-outline temperature might be an alert sign of a couple of infirmities such as injury and sepsis. The recording of the temperature extra through-out the long-term assists clinical specialists withdrawing in ends around the nation of wellness of the impacted individual with numerous infections. The to-be-expected spot approach of taking temperature is through method for the utilization of a temperature thermometer found in the mouth, ear, or rectum. Nonetheless, awful impacted individual relief and the exorbitant likelihood of pollution are typically an issue with those strate-gies. Be that as it may, the latest things in IoT- based innovation have advised various responses.

Some wearable devices were proposed, but they might be damaging the ear by produc-ing heat. It reduces the hearing and induces pain after some time. The infrared device helps to reduce the temperature inside the eardrum. The apparatus has a Wi-Fi sensor module and a record handling unit. Here, the deliberate temperature isn't provoked through the method of the environmental factors or diverse substantial exercises. The patient data are stored on the web page and can be accessed through the mobile phone or computer. In another review, a movable lightweight sensor was used to measure the child in real time. It additionally can alarm the father and mother assuming that the temperature transcends a fundamental level.

8.3.2.4 Blood Pressure Monitoring

Blood pressure measurement is the most important thing in the diagnostic procedure. Do the blood pressure measurement to register at least one person. The IoT technologies frequently monitor the temperature level and transfer it to the cloud. Blood pressure clas-sification both systolic and diastolic discussed. More than 100 people used it to predict the accuracy level, and it was validated. Design the model or device for the human body real-time monitoring for upcoming reference. The CNN-based machine learning model is used to classify the various levels of systolic and diastolic blood pressure assessment. ECG signals such as a photoplethysmogram can be measured with fingerprint. Oxygen satura-tion monitoring, glucose level monitoring. The microcontroller is used to read the value from the human body and the data sent to cloud storage.

8.3.2.5 Measuring Oxygen Saturation

Oxygen saturation describes the level of oxygen in your blood. The measurement is taken using a pulse oximeter; it is fixed on fingers. This is the easy way to quickly find the oxygen level. Fix it on four fingers and switch on the device; the oximeter reads the value and converts it into a signal. Lung conditions will have oxygen saturation between 88% and 92%, and it is clearly written in this chapter. The doctor or lung nurse otherwise assumes that normal is 96% or higher. The limitations of the oximeter are the absence of a good pulse and blood flow in fingers, cold hands, dehydration, or dirty probe.

The new method eliminates the error rate and provides day-to-day monitoring. The latest oximeter has the integration of IoT technology with cloud computing. It is a massive support for the healthcare industry. Oxygen saturation below 90% even without symptoms requires oxygen therapy. The level for neonates is 0.5 L/minutes, for infants or older children 1 L/2 minutes, for preschool 4 L max, and for adults 5 L Max. The non-invasive device would calculate the blood pressure along with heart rate and pulse. The read data are sent to the local server through Wi-Fi; in the web server, the data are received by GPRS. An alert system is developed to inform the patient and family members when the pressure level goes high or low.

8.3.2.6 Measuring and Monitoring Asthma

Asthma can affect the airway and difficulty in breathing, shrink of airway swelling of the air decreased. It may cause health issues such as chest pain, wheezing, and shortness of breathing. Any time, an asthma attack can happen and the solution to this is using an inhaler or nebulizer. Since there is a necessity for real-time monitoring, many asthma monitoring IoT systems have been proposed. By using a smart sensor, record respiratory rate and health numeric data stored in a cloud server for diagnostic purposes. Raji discussed alarm systems that use LM35 sensors to measure the respiratory rate. The device is proposed for an inhaler and exhaling air.

Data of respiration were sent to the nearby health center. A call is automatically sent to the patient once the threshold value is reached. Proper meditation can reduce asthma, and do not inhale very cold air. Smoking affects the lungs, and the system monitors the climate condition and directs the people to move from one place to another place. Machine learning algorithms also integrate IoT technologies and clearly track asthma. In a further development, many features have been added to monitor asthma.

8.3.2.7 Monitoring the Mood

Disposition checking presents extremely fundamental records roughly an individual's sentiments and is utilized to keep a restorative scholarly realm. It furthermore empowers experts the cure of several scholarly ailments love wretchedness, pressure, bipolar turmoil, and so forth the passionate realm further develops an individual's mastery in their scholarly realm. In an attitude mining approach, the CNN people group is utilized to examine and legitimize an individual's attitude, which might be articulated in half of 12 classifications: cheerful, energized, miserable, quiet, upset, and irate. In an identical review, the span of temper misuse became reached with an intuitive contraption called "Meezaj".

The application further affirms the meaning of bliss in better mental techniques and empowers the policymaker to separate the fundamental components that play an essential role in an individual's joy interaction. With the combination of a convoluted gadget getting to know calculation, pressure is presently identified before with the coronary pulse.

Moreover, the contraption can speak with the impacted individual around their strain realm. It is enchanting to state that the deformation state analysis might be helpful in making arrangements for an IoT- based contraption that might save you from an accident.

8.3.2.8 Management of Medication

In the health industry, drug management is a general issue. Non-adherence suits elder people and develops clinical conditions. In the keep going a few dissects focused on deciding the consistency of the patients with the medicine through the IoT gadget. Smart medical boxes are designed to remind people to take their tablets regularly. It has three compartments, in which every tray holds the medicine for three totally various hours: morning, noon, and night. The system by and large estimates various significant well-being boundaries (blood glucose level, blood gas level, electrocardiogram, etc.).

All recorded outcomes are saved to the cloud server. Doctors and patients access the enlisted data by abusing the portable application. In another review, information on the capacity status of comparative medications at temperature and stickiness was gathered together. Patients are upheld to keep up with the predetermined stockpiling climate. IVF strategy requires a thorough prescription program, so gadgets arranged will empower them to report their medications and report day-by-day infusions. They help track drug use history and contact well-being providers. In expansion, it was supposed that the Associate in Nursing adjusted the great IoT-based prescription framework to utilize the proper rationale to look at the data accumulated by the temperature sensor. The system is continuously monitoring the admitted patient body condition and temperature and automatically changes the time and dose of treatment.

8.3.2.9 Management of Wheelchair

A wheelchair is a part of a disabled person's life and limited mobility. It offers both physical and mental help. Nonetheless, from now on, the utilization of a wheelchair is limited because of mind harm. Henceforth, another investigation works on a bunch of activities, the route, and the global positioning framework with these wheelchairs. The IoT-based directing framework was proposed, which is coordinated with a period stretch impediment dismissal framework. The guiding framework perceives deterrents using picture handling strategies on the recordings of the recorded period. The utilization of mobile computing has made the seat on the board for patients exceptionally dynamic and simple. Through the blend of various sensors, portable innovations, and distributed computing, a keen wheelchair was created like the one painted in.

The framework incorporates a portable application that makes it more straightforward for patients to cooperate with the guardian in the wheelchair. The application likewise empowers parental figures to notice the good ways. In other review, an IoT-based wheelchair perception framework was created, which controls the seat with hand signals. The planned model is entirely reasonable for patients with tetraplegia. The hand motion data were recorded as a survivor of the RF gadget, which was a gift in the wheelchair. More sensor information was communicated to the server and put away in the cloud. Specialists/medical caretakers can get to the data from the cloud and utilize these data for analysis. It is important that in countless progressed and consequently delicate wheelchairs were reputed that observed seat development, yet additionally related umbrella, mat identification capacities for the head and obstructions. The planned framework offered a ton of monetary association with the living climate.

8.3.2.10 Rehabilitation System

A restoration framework was suggested, which utilizes a multimodal sensor to screen the patient's strolling design and assess development measurements. At the point when a patient utilized the smart walker, he/she estimates different development lattices, for example direction point, tallness, and strength. Specialists utilized the portable application to get to this information and create indicative reports, and a stroke recovery framework was created with a wearable wristband, mechanical hand, and AI calculation. The wrist splint was created with an IoT-based low-power material terminal; this sends the signal to the framework. Moreover, the 2D printed mechanical arm device was examined in muscle. Another review announced a games restoration framework that screens the temperature, practice the pose, electromyography, electrocardiography, and gave input to the athlete's players. Health experts would use the recorded data to anticipate patient recuperation and form recovery programs.

8.3.2.11 Other Notable Applications

One more potential application is the recognition of carcinoma exploitation various progressed AI calculations with a partner degree IoT-based framework. What's more, an ongoing investigation has conjointly educated the discovery of skin sores utilizing an IoT framework. Inside the style of ensuing age careful training place, the gadget utilized PC games to foster a preparation environment and it additionally furnished a stage to act with various specialists from better places. In a helpful human-robot framework fit for productivity doing expressions, a negligibly intrusive medical procedure was proposed. Utilizing a movable gadget, it's able to screen the degree of hemoglobin inside the blood. The gadget utilizes photoplethysmography sensors, a small weight radiating diode, and photodiodes to measure hemoglobin. This adequacy of the gadget has been extra-substantial by contrasting the outcomes and the setup of a colorimetric test.

8.4 Limitations, Challenges, and Opportunities

The healthcare industry has seen tremendous developments in the last 10 years, solving health-related problems. It has impressively further developed the medical care benefits, which have been presented at the fingertip. IoT has revolutionized the healthcare industry because of sensors, cloud computing, and technologies. With various advancements, IoT conjointly has bound difficulties and issues that give an expected extension for future examination. Some of the issues are referenced inside the resultant segment.

8.4.1 Servicing and Maintenance Cost

There have been fast advances in innovation as of late that would require persistent refreshing of HIoT-based gadgets occasionally. Each IoT-based framework comprises an enormous number of organized clinical gadgets and sensors. This implies high support, administration, and overhaul costs, which can put a strain on the organization's funds, and yet in addition on the end user, for which it is important to fuse sensors that can be worked with lower upkeep costs.

8.4.2 Power Usage

All the IoT devices run on a small lithium-ion battery. Suppose a power failure happens in the system; it is not so easy to replace the battery. We need to use either a high-capacity battery in the device, or a device that automatically regenerates power. The renewable energy system must integrate with all IoT devices.

8.4.3 Standardization

The majority of these products proclaim to notice the quality guidelines and conventions inside the style cycle but are invalid which is presumably normalized on these – essentially based absolutely HIoT devices for discussion conventions, data conglomeration, and dish interfaces. The approval and normalization of the advanced impacted individual records (EMR) recorded through the HIoT contraptions need to likewise be considered of their sum. This might be accomplished if several organizations and normalization of our bodies are appreciated, which incorporates IETF, ETSI, and net protocol for sensible objects. The artistic creations worked with specialists to make running gatherings to normalize the hardware.

8.4.4 Data Privacy and Security

Distributed computing has the concept of checking coordination frequently. However, this has conjointly made tending organizations a great deal of helpless to cyber-attacks. Postulation might cause misuse of patients' important information and will have an impact on the strategy for treatment. To thwart a HIoT framework from this vindictive assault, numerous preventive measures ought to be taken while thinking of a framework. The clinical and detecting gadgets encased in a HIoT network should evaluate and utilize personality confirmation, secure booting, adaptation to non-critical failure, approval of the board, whitelisting, watchword encryption, and secure matching conventions to stay away from an assault. Additionally, the organization conventions such as Wi-Fi, Bluetooth, and Zigbee, then, at that point, ought to be coordinated with got steering systems and message respectability check strategies. Since IoT might be an associated network where each client is joined to the cloud, any issue that happens inside the administration of IoT could think twice about the protection of the patient. This can be solved by developing a new algorithm in security and preventing unauthorized access to patient information.

8.4.5 Scalability

Versatility addresses a consideration instrument which could adjust to the changes inside the climate. A machine with better qualifications works great with next to no put-off and utilizes the available assets. Henceforth, it is fundamental to design a gadget with better versatility. The additional makes a machine additional green for present and fate employment. A HIoT machine is the interconnection of various logical devices, sensors, and actuators, which are acclimated rate data by means of the Internet. The deficiency of consistency in different related contraptions of a HIoT machine diminishes the quality of the machine and thus should be controlled speedily.

8.4.6 Identification

Medical services experts fight with different patients and patient caretakers at a comparable time. Suppose the patient has many health issues; they may consult with many doctors.

To know the previous health records, the doctor may be asked for the unique ID, so the patient and caretaker must know that log-in credentials. To avoid confusion about the patient records, the technology must be implemented in a good way.

8.4.7 Self-/Automatic Configuration

IoT devices provide more features such as automatic configuration and manual configuration. So only the user may change the parameter and they use it according to their needs. The user manual provides the user with the clear specification.

8.4.8 Continuous Monitoring

Numerous medical services circumstances call for the long-lasting period following the impacted individual for the span of cure as the instance of steady illnesses, coronary heart infections. In such cases, the IoT device should be fit for doing real-time follow-up productively.

8.4.9 Investigation of New Diseases

With the fast growth in the health-related industry, the mobile application has been developed and it is living with the humans and monitors the everyday activities. Many mobile applications are available in the Play Store, even though only a few applications work well only for particular diseases. Since some more features have been added to that application and improved the communication speed about the patient records.

8.4.10 Impact on Environment

Improving a HIoT, the machine requires mixing dissimilar medical sensors with rich semiconductors. Fabrication in most cases involves the use of ground steel and different toxic chemicals. It can also create a destructive impact on the environment. There is a need to create an adequate regulatory framework to manipulate and modify the production of sensors. In addition, more in-depth studies are needed to ensure that sensors use biodegradable materials.

8.5 Conclusions

This chapter looks at a couple of parts of the HIoT structure. This record covers an appreciation of the designing of a HIoT structure, its parts, and the correspondence between these parts. Likewise, this record gives information on current clinical consideration organizations where IoT-based advances have been explored. Using these thoughts, IoT advancements have helped clinical care providers screen and dissect different clinical issues, measure various prosperity limits, and give characteristic workplaces in faraway regions. This changed the region of the clinical benefits from a center structure to an all the more peaceful centered system. We, in a like manner, discussed the various HIoT system applications and their new examples. Likewise, the hardships and issues related to the arrangement, collection, and use of the HIoT structure were presented. Complete groundbreaking data on the HIoT devices have been obliged to the scientist who will begin their assessment just to make research progress in this field.

References

[1] Z. Ali, M. S. Hossain, G. Muhammad, and A. K. Sangaiah, An intelligent healthcare system for detection and classification to discriminate vocal fold disorders, *Future Generation Computer Systems*, 85, 19–28, 2018.

[2] G. Yang, L. Xie, M. Mantysalo et al., A health-IoT platform based on the integration of intelligent packaging, unobtrusive bio-sensor, and intelligent medicine box, *IEEE Transactions on Industrial Informatics*, 10, 2180–2191, 2014.

[3] Y. Yan, A home-based health information acquisition system, *Health Information Science and Systems*, 1, 1–12, 2013.

[4] A. Prasanth, S. Jayachitra, A novel multi-objective optimization strategy for enhancing quality of service in IoT-enabled WSN applications, *Peer-to-Peer Networking and Applications*, 13, 1905–1920, 2020.

[5] P. J. Nachankar, IOT in agriculture, *Decision Making*, 1, 3, 2018.

[6] V. G. Menon, An IoT-enabled intelligent automobile sys- tem for smart cities, *Internet of Rings*, 100213, 2020.

[7] E. Qin, Cloud computing and the internet of things: technology innovation in automobile service, in *Proceedings of the International Conference on Human Interface and the Management of Information*, pp. 173–180, Las Vegas, NV, USA, July 2013.

[8] K.B. Bhaskar, A. Prasanth, P. Saranya, An energy-efficient blockchain approach for secure communication in IoT-enabled electric vehicles, *International Journal of Communication System*, 35, e5189:1–25, 2022.

[9] D. Rajesh Kumar, K. Lalitha, Black-hole attack mitigation in medical sensor networks using the enhanced gravitational search algorithm, *International Journal of Uncertainty, Fuzziness and Knowledge-Based Systems*, 29, 297–315, 2021.

[10] S. Lavanya, A. Prasanth, S. Jayachitra, A tuned classification approach for efficient heterogeneous fault diagnosis in IoT-enabled WSN applications, *Measurement*, 183, 109771:1–22, 2021.

[11] H. Peng, Y. Tian, J. Kurths, L. Li, Y. Yang, D. Wang, Secure and energy-efficient data transmission system based on chaotic compressive sensing in body-to-body networks, *IEEE Transactions on Biomedical Circuits and Systems*, 11, 3, 558–573, 2017.

[12] J. Sekar, P. Aruchamy, An efficient clinical support system for heart disease prediction using TANFIS classifier, *Computational Intelligence*, 38, 610–640, 2022.

[13] L. M. Dang, M. J. Piran, D. Han, K. Min, H. Moon, A survey on internet of things and cloud computing for healthcare, *Electronics*, 8, 7, 768, 2019.

[14] B. Oryema, Design and implementation of an interoperable messaging system for IoT healthcare services, in *Proceedings of the 2017 14th IEEE Annual Consumer Communications & Networking Conference (CCNC)*, pp. 45–52, Las Vegas, NV, USA, January 2017.

[15] A. Ahad, M. Tahir, K.-L. A. Yau, 5G-based smart healthcare network: Architecture, taxonomy, challenges and future research directions, *IEEE Access*, 7, 100747–100762, 2019.

[16] M. N. Birje, S. S. Hanji, Internet of things based distributed healthcare systems: A review, *Journal of Data, Information and Management*, 2, 2020.

[17] K. T. Kadhim, An overview of patient's health status monitoring system based on internet of things (IoT), *Wireless Personal Communications*, 114, pp. 1–28, 2020.

[18] Y. Yuehong, The internet of things in healthcare: An overview, *Journal of Industrial Information Integration*, 1, 3–13, 2016.

[19] G. Shanmugasundaram, G. Sankarikaarguzhali, An investigation on IoT healthcare analytics, *International Journal of Information Engineering and Electronic Business*, 9, 2, 11, 2017.

[20] J.-Y. Lee, R. A. Scholtz, Ranging in a dense multipath environment using an UWB radio link, *IEEE Journal on Selected Areas in Communications*, 20, 1677–1683, 2002.

[21] H. Aftab, K. Gilani, J. Lee, L. Nkenyereye, S. Jeong, J. Song, Analysis of identifiers in IoT platforms, *Digital Communications and Networks*, 6, 3, 333–340, 2020.

[22] G. Cerruela Garcıa, I. Luque Ruiz, M. Gomez-Nieto, State of the art, trends and future of bluetooth low energy, near field communication and visible light communication in the development of smart cities, *Sensors*, 16, 11, 1968, 2016.

[23] R. Peng, M. L. Sichitiu, Angle of arrival localization for wireless sensor networks, in *Proceedings of the 2006 3rd Annual IEEE Communications Society on Sensor and Ad Hoc Communications and Networks*, pp. 374–382, Reston, Virginia, September 2006.

[24] D. P. Young, Ultra-wideband (UWB) transmitter location using time difference of arrival (TDOA) techniques, in *Proceedings of the Re Rrity-Seventh Asilomar Conference on Signals, Systems & Computers*, pp. 1225–1229, Pacific Grove, CA, USA, November 2003.

[25] R. Zetik, UWB localization-active and passive approach [ultra-wideband radar], in *Proceedings of the 21st IEEE Instrumentation and Measurement Technology Conference (IEEE Cat. No. 04CH37510)*, pp. 1005–1009, Como, Italy, May 2004.

[26] R. J. Fontana, S. J. Gunderson, Ultra-wideband precision asset location system, in *Proceedings of the 2002 IEEE Conference on Ultra Wideband Systems and Technologies (IEEE Cat. No. 02EX580)*, pp. 147–150, Baltimore, MD, USA, May 2002.

[27] L. Syed, S. Jabeen, S. Manimala, A. Alsaeedi, Smart healthcare framework for ambient assisted living using IoMT and big data analytics techniques, *Future Generation Computer Systems*, 101, 136–151, 2019.

[28] G. Marques, R. Pitarma, An indoor monitoring system for ambient assisted living based on internet of things architecture, *International Journal of Environmental Research and Public Health*, 13, 11, 1152, 2016.

[29] A. Dohr, The internet of things for ambient assisted living, in *Proceedings of the 2010 Seventh International Conference on Information Technology: New Generations*, pp. 804–809, Las Vegas, NA, USA, April 2010.

[30] C. Tsirmpas, A. Anastasiou, P. Bountris, D. Koutsouris, A new method for profile generation in an internet of things environment: an application in ambient-assisted living, *IEEE Internet of Rings Journal*, 2, 6, 471–478, 2015.

[31] R. Maskeliunas, A review of Internet of Things technologies for ambient assisted living environments, *Future Internet*, 11, 259, 2019.

[32] M. S. Shahamabadi, A network mobility solution based on 6LoWPAN hospital wireless sensor network (NEMO-HWSN), in *Proceedings of the 2013 Seventh International Conference on Innovative Mobile and Internet Services in Ubiquitous Computing*, pp. 433–438, Taichung, Taiwan, July 2013.

[33] R. Tabish, A 3G/WiFi-enabled 6LoWPAN-based U-healthcare system for ubiquitous real-time monitoring and data logging, in *Proceedings of the 2nd Middle East Conference on Biomedical Engineering*, pp. 277–280, Doha, Qatar, February 2014.

[34] C. Sandeepa, An emergency situation detection system for ambient assisted living, in *Proceedings of the 2020 IEEE International Conference on Communications Workshops (ICC Workshops)*, pp. 1–6, Anchorage, AL, USA, June 2020.

[35] G. Marques, I. M. Pires, N. Miranda, R. Pitarma, Air quality monitoring using assistive robots for ambient assisted living and enhanced living environments through internet of things, *Electronics*, 8, 12, 1375, 2019.

[36] G. Marques, R. Pitarma, A cost-effective air quality supervision solution for enhanced living environments through the internet of things, *Electronics*, 8, 170, 2019.

[37] X. M. Zhang, N. Zhang, An open, secure and flexible platform based on internet of things and cloud computing for ambient aiding living and telemedicine, in *Proceedings of the 2011 International Conference on Computer and Management (CAMAN)*, pp. 1–4, Wuhan, China, May 2011.

[38] H. Mora, D. Gil, R. M. Terol, J. Azorın, J. Szymanski, An IoT-based computational framework for healthcare monitoring in mobile environments, *Sensors*, 17, 10, 2302, 2017.

[39] S. Tyagi, A conceptual framework for IoT-based healthcare system using cloud computing, in *Proceedings of the 2016 6th International Conference-Cloud System and Big Data Engineering (Confluence)*, pp. 503–507, Noida, India, January 2016.

[40] S. Nazir, Internet of Things for Healthcare using effects of mobile computing: A systematic literature review, *Wireless Communications and Mobile Computing*, 2019, Article ID 5931315, 20 p., 2019.

[41] R. S. H. Istepanian, D. Casiglia, J. W. Gregory, Mobile health (m-Health) for diabetes management, *British Journal of Healthcare Management*, 23, 3, 102–108, 2017.

[42] L. Chuquimarca, Mobile IoT device for BPM monitoring people with heart problems, in *Proceedings of the 2020 International Conference on Electrical, Communication, and Computer Engineering (ICECCE)*, pp. 1–5, Istanbul, Turkey, June 2020.

[43] S. H. AlMotiri, Mobile health (m-health) system in the context of IoT, in *Proceedings of the 2016 IEEE 4th International Conference on Future Internet of Rings and Cloud Workshops (FiCloudW)*, pp. 39–42, Vienna, Austria, August 2016.

9

Case Studies: Cancer Prediction and
Diagnosis in the IoMT Environment

C. Soundaryaveni

Sri Krishna College of Engineering and Technology

A. Prasanth

Sri Venkateswara College of Engineering

S. Lavanya

Muthayammal Engineering College

K.K. Devi Sowndarya

DMI College of Engineering

CONTENTS

DOI: 10.1201/9781003256243-9

9.1 Introduction

Cancer has become one of the world's most prevalent diseases. Cancer has surpassed heart disease as the second most common incurable disease, according to WHO research. A total of 9.6 million people are impacted, with one out of every six deaths every day. Breast cancer incidence rates climbed dramatically in several countries, owing to a combination of risk factors and increased detection [1]. Cancer and death are caused by a single cell that grows incorrectly and uncontrollably. People aren't even aware of cancer in its early stages since cancer symptoms differ. Cancer symptoms are documented after a day, all cells are handled, and the symptoms that follow lead to cancer's final stage. According to the IoMT, technological improvements aid in the prevention and management of cancer. A multi-cancer early detection (MCED) test uses liquid biopsies (blood-based screening tests) to diagnose early cancer stages before symptoms appear. Lung, breast, cervical, colorectal, and prostate cancers are presently the only malignancies covered by this screening test. As a tool, intelligence can aid in disease detection and prediction in the future [2].

The Internet of things (IoT) has resulted in the exploration of artificial intelligence, which aids in the tracking of our entire bodily system's functionalities through sensors. Sensors will collect information from our bodies and store it in physical storage devices for health monitoring [3–6]. The data gathered from the body are then examined. The combination of IoMT with machine learning (ML) improves cancer early detection. Early detection and evaluation of disease causes are aided by ML technology [7]. With the use of supervised learning, the senior pathologist inspected it. There are millions of tissues and cell membranes in our bodies. Every cell in our body has a distinct characteristic that aids in the biological growth of our bodies. Cancer cells are formed as a result of mutations, the environment, microorganisms, lifestyle, and food. There are a variety of cancer symptoms and characteristics to choose from. Each illness has its own set of signs and symptoms. The size and location of the tumor, as well as the presence or absence of metastases, all have a role.

A human body is made up of millions of cells, each of which has nuclear membranes that can reconstruct dead cells. Cancerous cell proliferation is fueled by DNA (deoxyribonucleic acid) damage and unregulated cell expansion. If a cell lives for a longer time in the body, it will be manipulated. Throughout the human body, this will lead to the generation of cancer cells. Human cells are capable of self-replication and growth control in their natural state. It will go on unless human life comes to an end.

The microorganisms manipulate together in some organs or the bloodstream as a by-product of a biological transformation in the human body, eventually forming cancer cells. As a single cancer cell divides into two, it will divide into quadruples as it travels throughout the human body. The homologous cells are formed by the repeated process of genetic reduplication. This spreads quickly and destroys the mammalian body's natural healthy tissue. The cells that originate from them could seem dissimilar from healthy cells. Abnormal cells can cause damage to the skin, organs, and blood, as well as destroy neighboring tissues and cause malignancies to form. The tumor cell has no understanding of when it should stop reproducing and when it should die [8].

These cells will not adhere together, so they will break off and travel through to the blood vessels in numerous multiple directions eventually commencing to grow somewhere inside the body, a process known as metastasis. Among the most useful for distinguishing in our bodies is our immune system. They prevent the skin from deficiency and serve as a deterrent against alien bodies. Macrophages are specialized cells that identify bacteria and some other dangerous microorganisms. Moreover, they can present antigens toward T cells and intensity and frequency by releasing hormones that activate other cells. Immunity cells that combat cancerous cells include natural killer cells, methylation patterns suppressor cells, and dendritic cells. Since cancer cells grow outside of the immune system's control, they are unable to safeguard the healthy cells in our bodies. The image illustrates the differences between a normal cell, a cancer cell, and cancer cells that proliferate uncontrollably.

Oncology is the study of how to prevent, diagnose, and treat disease in humans. Tumors naturally generate numerous tumor cells, some of which are recoverable and many others can be evacuated. The majority of tumor cells can be found among both men and men and women's lungs, stomachs, and colons. Heterosexual men's tumors encompass prostate, urinary bladder, and melanoma, and female sexual tumors include breast, endometrial and uterine cancers, and thyroid cancer [9]. The four components of cancer are as described in the following: Stage Each: Cancer cells are in their original place in the in situ stages. Stage 1: Cancer cells can obtain passage into the plasma membrane, which would be a fibrous, thin roadblock that covers and protects the tissue where cancer developed. After cancerous cells have indeed been formed, they proliferate and generate damaged tissue. Tumor clinical manifestations: A tumor is a collection of symptomatology that could also fluctuate between moderate to severe. The sensations varied depending upon the organs impacted. After cancerous cells have indeed been formed, they proliferate and generate damaged tissue (Figure 9.1).

Tumor Clinical Manifestations: A tumor is a collection of symptomatology that could also fluctuate between moderate and severe. The sensations varied depending upon the organs impacted. Symptoms are caused by the cancerous cell pushing against circulation

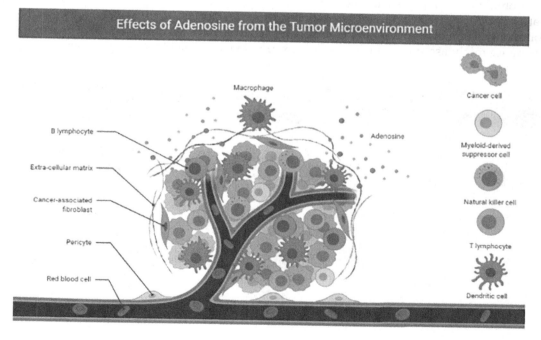

FIGURE 9.1
Effects of adenosine from the tumor microenvironment.

arteries, nerves, and organs. Among the most frequent symptoms of cancer are losing weight, temperature, discomfort, and exhaustion. Treatment chemotherapy aims to shrink tumors. The development of medical oncology is concerned with the detection, treatment, and prevention of cancer. Chemotherapy and hormone treatment are some of the treatments that have been analyzed for strengths and weaknesses, the physician's recommendation, and indeed the stage of cancer [10].

Cancerous cells inside the human body are affected by four different stages. They went through various cancer treatments to manage cancer symptoms and their detrimental impacts [11–14]. The purpose of radiation oncology is to use relatively high radio waves to treat damaged cancer cells. Adverse problems are associated with both of these treatments, but they differ from person to person. Surgical oncology removes tumors through operations, and a small portion of tissue is extracted for biopsy.

9.1.1 Types of Cancerous Tumors

In the early stages of the disease, symptoms are impossible to determine [15,16]. Clinical manifestations and withdrawal symptoms are identical only when they occur during forecast period 1 or Stage 2, but people see them at the end phases. The images above show a variety of tumors that affect the body's cells. The image above illustrates their preferences (Figure 9.2).

Melanoma is a type of skin cancer that affects millions of people worldwide. It appears as more than just a small dark dot on the skin, gradually grows, and distributes all over the body. This was the most hazardous and spread to other parts. Overexposure to the sun, light skin, and a family medical history of pigmentation all seem to be independent predictors. Cancer is a condition that would be difficult to control.

Glioblastoma is indeed a neurogenic malignancy. It is mainly composed of cell differentiation supporting nerve cells in the brain microglia. It impacts elderly people increasingly commonly. Hangovers, nausea, vomiting, and seizures have become worse because of it. Therapies might assist in reducing cancer progression and relieving the symptoms.

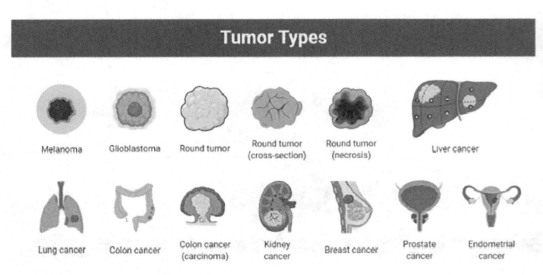

FIGURE 9.2
Tumor types.

Round Tumor: Round cells that have been squeezed and have a greater nuclear-cytoplasmic ratio. Cylindrical cellular stem cells derive either soft tissue or bones, and they most frequently damage adolescents and older children.

Liver Cancer: The liver is a professional soccer tissue found in the anterior part quadrant of the abdomen, underneath the esophagus, and above the stomach. The most prominent sort of cancerous cell is hepatocytes carcinoma.

Lung Cancer affects men and women similarly. Lung cancer would be most frequently diagnosed in adults aged 65 and above. Lung cancer is caused by both smoking and non-smoking, and the majority of the research consisted of DNA mutations.

Colon Tumor begins in the large intestine and subsequently spreads throughout the body. The colon is the conclusion and recommendations of the gastrointestinal tract. It affects people of any age, particularly seniors. It includes polyps, which seem to be tiny aggregates of cells that make up the inside of the colon. Their tumors might turn into hepatocellular carcinoma over time. Mutations in women's DNA cause their cell cycle to progress and reproduce fast, eventually leading to kidney cancer. The tumor's aberrant cells can migrate beyond the kidney, and maybe some cells can even travel to other systems of the brain. Symptoms are rare inside the early stages of the disease. Loss of appetite, decreased appetite, fatigue, fever, and blood in the urine that appears pink, red, or cola-colored are all symptoms to watch out for. Breast tumor is the second most commonly diagnosed cancer in women, after skin cancer. Cancer can strike both men and women, although women are far more likely to develop it.

Breast Tumors generate a lump or thickening in the surrounding tissue, which would be a sign of the tumor [16]. Dimpling, for example, is indeed a change in the skin of the breast. The skin over the shoulders is wrinkled or pitted. Some mammary cells develop inappropriately, culminating in this condition. Cells in the milk-producing ducts are the start of malignancy.

Lymphoma is passed down through the generations due to a family history of depression. To dramatically reduce risk factors, they should just get a blood test. Males have a small walnut-shaped gland called the prostate that generates seminal fluid, which nourishes and transfers sperm. They grow slowly and are restricted to the prostate gland, where they are unlikely to cause significant harm.

Prostate Cancer does not have any advanced warning or symptoms. Symptoms include difficulty peeing, decreased urine force, bone discomfort, and erectile dysfunction. This occurs as a result of fast cell development in specific glands or organs. Obese people may have an increased testosterone level than people who are regarded to be of normal weight, while studies have yielded inconsistent findings.

Endometrial Cancer is a subtype of uterine cancer that starts in the lining of the womb. It starts in the squamous epithelium that makes up the uterine lining. Endometrial carcinoma is also known as uterine cancer. Because endometrial cancer commonly causes irregular vaginal discharge, it is frequently detected at an early stage. Bleeding during periods and pelvic pain are common symptoms. This occurs as a result of aberrant cell proliferation.

9.1.2 Cogitation on Cancer Prediction

Cancer is the second strongest disease, impacting one out of every six people nearly every day. The reduplication of cancerous cells is highly influenced by cellular damage. The human immune system maintains us healthy and protects us from antibodies. There are really now more tumor studies and research becoming explored. Oncology is a branch of medicine concerned with the prevention of cancer. The sorts of aggressive diseases are used to characterize patients who are treated for cancer, and chemotherapy is offered to them depending on the specific stage. The emphasis of cancer prevention research has migrated from population epidemiology studies to finding higher-risk precancerous abnormalities in individuals utilizing developing early detection technology.

These medical-based chemopreventive techniques have the potential to significantly reduce the incidence of cancer globally. Therapeutic approaches are indeed being developed to inhibit the pathways that have been influenced by genetic and molecular changes that have led to cancer growth. In order to prevent the advancement of in vitro tumors to invasive cancer, treatments are aimed at preventing late-stage, but still pre-malignant, processes that lead to them. Vaccinations aid in the prevention of immune system harm and the maintenance of overall health. Immune response approaches that emphasize immunology, which aids in the good maintenance of good health, minimize the high-risk factors of tumor detection vaccination for human papillomavirus (HPV) and hepatitis B virus (HBV). Improving one's lifestyle and being fit can help to lessen the high-risk factors associated with immune system deterioration. Chemoprevention is a pharmacological strategy for cancer prevention that focuses on the transition from pre-malignant cancer to invasive malignancy. Chemoprevention effects on high-risk tissue prevent tumors from developing in healthy or pre-malignant tissues of cancer [17].

Chemopreventive phytochemicals can prevent multi-step carcinogenesis in its tracks. Antibiotics can stop malignant melanoma from turning into cancerous tissue. Drugs are injected into certain organs at different stages of cancer to suppress cell growth. Chemotherapy is not a cure for cancer; rather, it is a tool to help restrict tumor cell proliferation. Another doctor recommends radiotherapy as a treatment option. The maximum tolerable dose to nearby normal tissues limit the tissues. The treatment of radiotherapy took into account how to enhance the ability of cancer cells to kill themselves as well as the ability of nearby healthy tissues to withstand radiation damage. It doesn't kill cancer cells right away; it takes a day or a week of treatment for chronic cells' DNA to be destroyed enough for them to die.

After radiation therapy, the tumor continues to die for weeks or even months. Cell radiation divides treatment into exterior and internal beams based on the tumor. The tumor's size, type, and size are all assessed early in its development. External beam radiation therapy (EBRT) moves around the body, aiming radiation externally without touching it. Radiation can be directed in any direction to a region of the body. This treatment is focused on certain regions of the body, such as the lungs, where radiation is only applied to the chest and not the entire body. Internal radiation is a form of radiation treatment in which the radiation source is placed inside the body.

Radiation can come from a solid or liquid source. Brachytherapy is the term for solid-source radiation. Seeds, ribbons, or capsules containing the radiation source are implanted near the tumor cells inside the body in this method of treatment. It, like external radiation, exclusively affects certain regions of the body. Systemic therapy refers to the use of a liquid source of radiation, a medication that goes throughout the body's blood tissues, seeking out and killing cancer cells. To diagnose, a biopsy is performed.

A biopsy is a removal in which a pathologist extracts and examines a tiny piece of tissue under a microscope. Depending on the type of tumor detected, tissue could be extracted from the tumor. Tissue samples are performed by surgeons who remove a full organ or grow a therapeutic process based on the type of malignancy. Diagnostic tests of the second type will extract tumor specimens using an endoscope, which is a narrow needle. A needle biopsy can be done in two ways. Solid needle biopsies can be used during a lumbar puncture. A fine-needle biopsy involves extracting a little amount of distilled water and

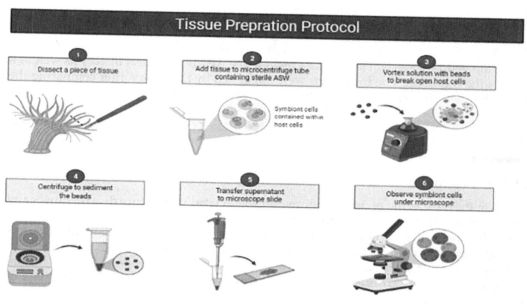

FIGURE 9.3
Tissue preparation protocol.

very small fragments of tissue from the tumor with a very thin needle linked to a syringe. If the tumor is located further within the body, the needle can be guided by an imaging test such as an ultrasound or CT scan. The advantage of FNA is that the diagnosis can be made on the same day as the test, and there is no need to remove tissue from the body (Figure 9.3).

- **Step 1** of the tissue preparation protocol is to remove a portion of tissue or fluids from a tumor in the body. Tumor cells can be found in any part of the body. If tumor cells cannot be removed, they will be treated with surgery inside the body. Surgery can be performed without any prior planning.
- **Step 2:** Continue sterile AWS while inserting tissues into the microcentrifuge tube. This test aids in a more in-depth investigation of cell functions while under sterile conditions. Steamy cells are where symbiotic cells live.
- **Step 3:** Centrifuge the beads to sediment them. Separating solids from liquids is a procedure. To recover solids or liquids from slurries, it uses the G-forces generated by high-speed spinning.
- **Step 4:** Observe the supernatant by transferring it to a microscope slide.
- **Step 5:** Use a microscope to examine symbiotic cells.

Biopsies aid in the reduction in tumor cells and the preservation of human life. The biopsy and therapy are variables depending on the type of tumor in the body. If they are in Stage 1 or Stage 2, cancer recovery can take quite a while. Evaluation of the rooster: A small sample of tissue or liquid from the body is removed and examined under a microscope. The senior pathologist is responsible for this testing. The science of pathology is the study of diseases and ways to prevent cell harm. Pathology test findings assist doctors in diagnosing disorders and prescribing appropriate therapies. Under microscopic analysis, the tissue preparation technique is depicted in the image below. On the testing plate, a small amount of liquid is inserted and analyzed. The results are compared to the same set of data that were evaluated previously. The test data include the patient's medical history as well as a study of the condition. The treatments were suggested by the doctor after comparing the history. Surgical or non-surgical therapy options are available.

9.2 ML and IoMT

ML aid in the forecasting of data or information associated with future disease causes and consequences [18]. Analyses for future forecasts are based on previously collected datasets from different entities. It facilitates the diagnosis and treatment of malignant disorders for people who undergo oncology and pathology. ML also aids in the discovery of several solutions to healthcare-related issues. Various topics, such as computer science, statistics, and optimization, are covered by machine learning techniques. Using example data or experience, computer programming is used to optimize a performance criterion. After a model has been created according to some characteristics, a computer software optimizes the criteria of the module training data as well as practice.

The module could be predictive, making future predictions, or descriptive, based on the data. In order to develop models, machine learning employs statistics theory, and the essential foundation is established by mathematical inference. Machine learning serves two purposes: The first is training, and the second is knowledge. (a) Training: A strong algorithm is used to solve the optimization problem while also storing and processing a large amount of data. (b) Learned: representation and algorithmic solution for inference needs. The speed and time complexity of the classification step are just as significant, including its predicted accuracy.

Almost all ML issues may be expressed as optimization problems for different types of clustering sets at their core. There are three forms of machine learning, and the majority of applications are tested and taught in categories 1 of 3: supervised, unsupervised, or reinforcement learning [19]. They have not yet implemented a machine learning method in the existing system. As input, the datasets are entered into the computer and tested. Without even any predefined data, various algorithmic approaches could be used to incorporate the testing data. As a result, it calculates the input in terms of values and forecasts the output in terms of reports. Those reports will include points and measures that are accurate, but data without a practical strategy take time to treat. The method is time-consuming, and it may cause therapy to be delayed or result in days of monitoring. Considering supervised learning and unsupervised learning, the best way to find solutions to challenges and decision-making processes on the network of healthcare things is to use machine learning approaches. In general, determining the data assessment, goal definition, and the available review algorithm are a good method for a perfect machine learning approach. (a) Data evaluation: Determine whether or not labels are included in the dataset, and whether or not the column is labeled or unlabeled. This will assist in determining if the dataset should be defined using a supervised, unsupervised, semi-supervised, or reinforced learning method. (b) Establish a goal: If the situation is recurrent, what would the intention be or a new problem that the algorithm is meant to solve. The purpose is to determine the data collection that will be used to test the prediction condition. (c) Reviewing the algorithm available.

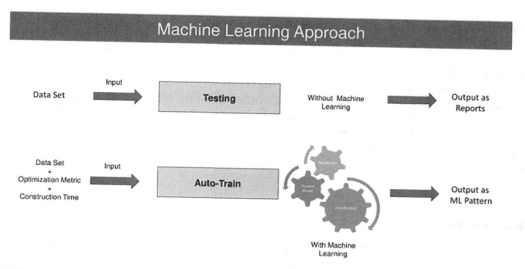

FIGURE 9.4
Machine learning approach.

The algorithm should be appropriate for the overall amount and structure of the data. It's possible that the issue is related to the dimensionality of several futures, traits, or characteristics data. The future prediction datasets are defined by these three parameters. Both datasets may contain an empty or null value, which should have been recognized before beginning the forecast. Supervised and unsupervised learning are critical concepts in machine learning that must be grasped before implementation (Figure 9.4).

Information is collected as data collection and processed by a computer using optimization parameters in the suggested system. The settings are set using an algorithm that aids in the measurement of decision-making metrics' speed and correctness. The speed determines how fast it may spread throughout the body, while the accuracy determines its boundaries. The efficiency of the input dataset is calculated using construction time. The data are automatically calculated by the auto-train with machine learning using algorithms that measure the dataset, optimization metrics, and building time. In ML, the two main learning approaches are classification and regression [20]. The process of organizing the data into several category classifications, which can be separated into binary or discrete labels, is known as classification.

The data are labeled based on numerous parameters in the input, and the labels are anticipated for the data. The "if-then" mapping relationship rules represent the classification connection. The values are mapped to predefined classes using the classification's map function. Classification employs algorithms such as decision tree (DT) and logistic regression (LR). The process of determining a link or function in a model is known as regression. Based on the historical dataset, dispersion movement may also be determined. In regression, the mapping function is used to convert values into a continuous output. Linear regression is utilized in the regression of algorithms such as the random forest. A ML pattern is the outcome of classification and regression. The pattern can be organized into graphs or charts, which are visual representations of datasets. Based on the data collection, machine learning is classified into two types: supervised and unsupervised learning. Datasets are well trained and integrated in the computer in the training set, and prediction becomes simple with a predetermined dataset.

Unsupervised learning algorithms figure out for themselves what makes a dataset unique or interesting. The computer must be able to program itself and analyze both structured and unstructured input to show its existence. There are two types of unsupervised machine learning: anomaly detection and clustering. Clustering seems to be an example of unsupervised machine learning that is widely used. Clustering is the method for organizing unlabeled data into groupings called clusters, which entails locating data fields that are similar across a data collection. Finding or detecting unusual things or events that differ considerably from the rest of the data is what anomaly detection is all about. When a data anomaly is suspicious, it is frequently used to discover bank fraud and medical errors.

9.3 Best Cancer Prediction and Diagnosis Using Supervised Algorithm

In supervised learning, the model is "trained" with a large amount of data, and it is trained or defined by its ability to reliably categorize data or predict outcomes using labeled datasets [21]. These models are trained on the labeled dataset and used to forecast future events. Breast tumors are one of the frequent common female cancers, and they can spread rapidly in females; it is the main cause of death. It's important to realize that the vast majority of breast lumps aren't malignant. They aren't cancers, and the abnormal growth in the breast doesn't spread. Fibrocystic breasts can raise a woman's chances of developing a tumor. Fatty tissue makes up the majority of this organ. Certain abnormal growth in the mammary can develop into a tumor, which must be assessed by a medical professional to identify whether it would be benign (non-cancerous). Breast cancer can form in separate parts of the breast, a muscular organ that sits on top of the upper ribs and contracts when muscles engage.

The preponderance of this organ is made up of glands, ducts, and fatty tissue. A trained dataset is the input, and the learning algorithm generates predictions about certain unique, unseen observations that the model could be given. While using the procedure to train, the mapping function from the model input dataset has the input variable "X" and the expected output variable "Y".

$$'Y.'Y = f(X) \tag{9.1}$$

When fresh data (X) can forecast, the output variable (Y) for the data, the purpose is to discover the approximation mapping function. The expected output of input "X" is then given by "Y". The two most common forms of supervised models are classification and regression. The statistical process of teaching a computer to perform operations on a given dataset is known as automation. The practice of classifying data collection into classifications that can predict sets of data in both structured and unstructured ways is known as classification. The target, label, or categories were represented by the classifications. Resembling the mapping function from discrete variable inputs to discrete variable outputs is the problem of predictive modeling. Using classification algorithms performed on the training data to detect the same pattern in subsequent sets of data, the result is a sort of "pattern recognition." Depending on the dataset, multiple types of categorization methods are applied.

The LR, naive Bayes, k-NN, DT, and support vector machine (SVM) are the most common algorithms utilized in ML [22–24]. In machine learning, there are various types of categorizations predicting models. They are divided into five categories: predictive modeling, binary classification, multi-class classification, multi-label classification, and imbalance classification. The classification predictive model aids in the categorization of labels in a data collection for the prediction of the future. Assigning a class label to input instances is part of the classification prediction model. Multi-class classification predicts one or more of two classifications, while binary classification predicts one of two classes. In multi-label classification, each example is assigned to one or more classes, while imbalance classification refers to tasks in which the distribution of examples across classes is not equal (Figure 9.5).

The tumor cell at Stage T1 is quite tiny, measuring 2 cm or less in diameter. T2 and T3 have a larger size, ranging from 2 to 5 cm wide, and have expanded to tissues near the breast. Any size tumor can develop into the chest wall. T4 is the last stage of tumor cell proliferation, and it causes the breast to shrink. A tumor is formed by a single aberrant proliferation of cells; it can be avoided in its early stages if it is identical. The tumor can spread throughout the body when a cancer cell enters the bloodstream or lymphatic system and is spread to many areas of the body. A lymph node is a part of the immune system that transfers lymph fluid away from the breast, allowing cancer cells to enter lymph veins and spread to lymph nodes. Most lymph veins in the breast pass through lymph nodes beneath

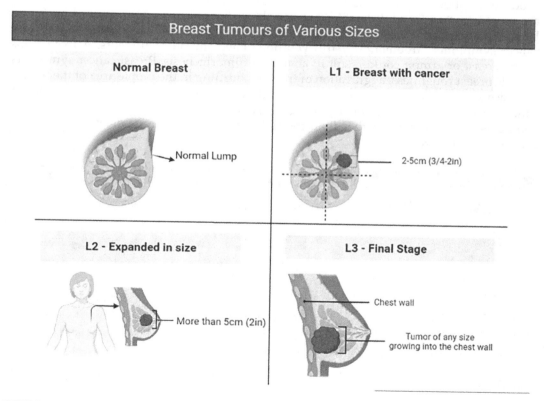

FIGURE 9.5
Different breast tumor sizes. (For research purposes, use BioRender as a source.)

the arms, inside the chest around the breastbone, around the collar bone, and below the collar bone. If, indeed, the cancer cell spreads, through the lymphatic system, it has a larger chance of spreading to other parts of the body.

Breast cancer can be triggered by inherited gene mutations or a family history of the disease. As an adult, increasing body weight puts you at a higher risk of breast cancer. It is more vital to balance diet and intake with physical exercise in order to live a healthy life. The cells in the breast that can become cancerous determine cancer. Typically, cancer cells start in the epithelial cells of the body. In situ and invasive are the two forms of breast cancer. Tumor that begins in situ begins in a milk duct, which is a tube with well-defined walls that hasn't expanded into the other breast wall. Tumor that has spread to the surrounding breast tissue is known as invasive ductal cancer. Triple-negative breast cancer cells lack estrogen and progestin, as well as the ability to produce protein.

9.4 Early Detection of Tumor Cells and Symptoms of Breast Cancer

The breast takes on a familiar aspect as tumor cells proliferate. Rapid cell growth causes a proliferation of abnormal cells, which accumulate over time to form a lump or mass. The ducts, a tube that delivers lactic fluid to the lobules, the glandular structures that produce lactation, are where the majority of breast tumors develop. Breast tumors that are more common in adult females are now being identified in men. According to the latest statistics, about 2,600 men will be diagnosed with breast cancer this year. Breast cancer symptoms can emerge even if you don't have any other ailments. A fresh mass in the breast or armpit could result in abnormal tumorigenesis. Breast cancer symptoms include breast roughness, aggravation of the skin, burning in the nipple area of the breast, decrease in breast size, and soreness in the milk ducts. The most common benign breast cancers are vesicles, high levels of estrogen, glandular disorder, canal, villoma, and fat sphacelus. The left breast, rather than the right, is where the large percentage of breast cancers start (Figure 9.6).

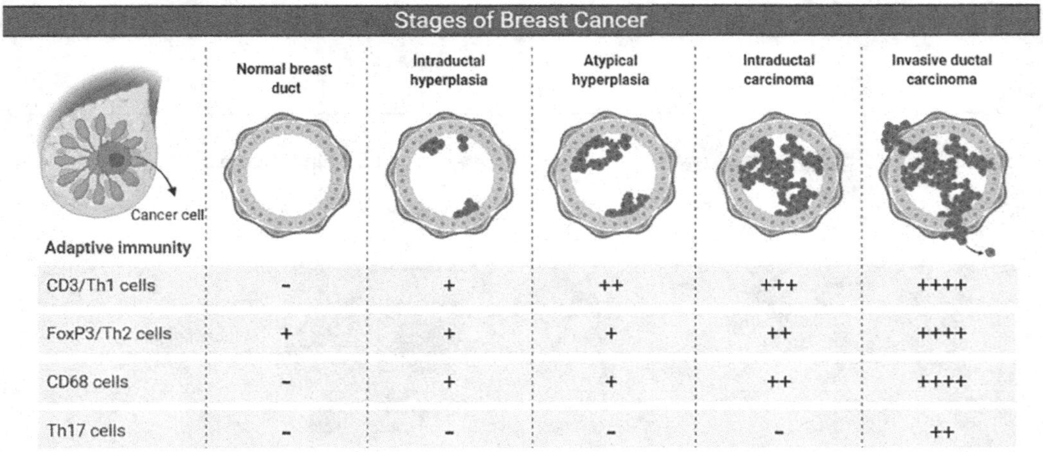

Stages of Breast Cancer					
	Normal breast duct	Intraductal hyperplasia	Atypical hyperplasia	Intraductal carcinoma	Invasive ductal carcinoma
Adaptive immunity					
CD3/Th1 cells	–	+	++	+++	++++
FoxP3/Th2 cells	+	+	+	++	++++
CD68 cells	–	+	+	++	++++
Th17 cells	–	–	–	–	++

FIGURE 9.6
Stages of breast cancer. (For research purposes, use BioRender as a source.)

The immune system is a complex network that defends the organ from sickness and endothelial dysfunction. It recognizes and responds to a wide range of viruses and pathogens in the body, as well as cancer cells, and attempts to combat abnormal cell growth. The body's defense system fights foreign bodies stuck inside tissue organisms as well as healthy tissue. Innate and adaptive immune systems are two subsystems of the immunity system. The innate immune system responds to a wide range of foreign substances with a pre-programmed response. By learning to detect molecules that have already been encountered, the adaptive immune system can deliver a tailored response to each alien body. The graphic above depicts the aberrant development of cells at various stages. Adaptive immunity systems monitor cell growth and try to comprehend rapid changes in cell size. CD3/Th1 cells, FoxP3/Th2 cells, CD68, and TH17 cadre are adaptive phagocytes that keep track of the body's biological changes.

9.5 Various ML-Based Breast Cancer Classification

9.5.1 LR Classifier

The regression learning model uses a set of attribute variables to predict data-dependent qualities. In ML, two types of regression analysis techniques are utilized to tackle regression problems: linear regression and logistic regression. In binary logistic regression, there are just two choices available: "death" or "alive." Because this allows for the easiest interpretation, the outcome is often coded as "0" or "1". If the possible outcome is a success, it is marked as "1"; defeat is marked as "0", whereas success is marked as "1". Based on the values of the autarkic variables, logistic regression has been put upon to forecast the chances of predictors. To estimate the probabilities, the likelihood of a given outcome becoming circumstance has been multiplied by the possibility of it being a non-case.

LR is utilized to model the probability of classification issues with two alternative solutions. The aim is to test inputs as accurately as possible with the lowest potential error rate. To evaluate the accuracy and efficacy of various approaches, the correctness and development time approach is employed. A statistical strategy for predicting binary classes is logistic regression. The result is a categorical variable, also termed as the target variable. The word "bipolar" alludes to the fact that there are only two possible groups of people with bipolar disorder. It could be used to help in cancer detection, for example. It calculates the likelihood of an event occurring with a categorical target variable, of linear regression. The dependent variable is the log of odds. The logit function is used in logistic regression to predict the likelihood of a binary event occurring.

LR produces a static result. To compute logistic regression, the classifier evaluation (ML) method is employed. Since logistic regression is a linear classifier, the linear function $f(x)=b_0+b_1x_1+\ldots+b_rx_r$, frequently referred to as the maximum likelihood estimation. The regression coefficient predictions, also known as predicted scores or coefficients, are represented by the variables $b_0, b_1, \ldots, b_r$.

$$p(x)=1/\left(1+f(x)\right) \tag{9.2}$$

where f intends the logistical function (x). The result is frequently near 0 or 1. The p type is generally used to express the likelihood that the output for a given x equals 1. As a result, the probability of an output of 0 is $1\ p\ (x)$. The optimal expected values $b_0, b_1..., b_n$ are determined using logistic regression, and the value $p(x)$ is as close to the actual responses as feasible. $y_i, I=1,..., N$, where N is the number of occurrences. Training process T is to find the best weights based on available data. The log-likelihood function (LLF) is often maximized for all occurrences to obtain the appropriate weights. $n=1,..., N$.

$$\text{LLF} = i(y_i \log(p(xi)) + (1y_i)\log(1p(x_i))) \tag{9.3}$$

The approach, as maximum likelihood estimation, is shown by the computation. The LLF for the related information is identical to log $(1\ p(x_i))$ when $y_i=0$. If $p(x_i)$ becomes close to $y_i=0$, log $(1\ p(x_i))$ gets close to 0 – this is the desired outcome. While $p(x_i)$ is far from 0, log $(1\ p(x_i))$ decreases significantly. Because you want as much LLF as possible, you don't want that to happen. When $y_i=1$, the LLF for that observation is $y_i \log(p(x_i))$. If $p(x_i)$ approaches $y_i=1$, $\log(p(x_i))$ approaches 0. $\log(p(x_i))$ is a very large negative number.

9.5.2 SVM in Cancer Detection

SVM is a supervised machine learning algorithm that is often used to handle binary classification problems. It learns by being fed a dataset. Establishing a straight line between two classes is how a simple linear SVM classifier works. That is, most of the data points along one side of the line appear to be classified into one category, while data points on the other side appear to be assigned to a different group. This implies that there may be an endless number of lines from which to choose. The linear SVM algorithm is superior to other algorithms such as k-nearest neighbors because it selects the best line to classify your data points. It picks the line that divides the data and is the farthest distance from the closest data points. The algorithm finds a hyperplane (or decision boundary) using this dataset, which should ideally have the following properties:

- It generates a maximum margin of separation between samples of two classes.
- Its equation $(x'+b=0)$ returns a score of 1 in positive situations and 0 in negative situations.

The equation for a hyperplane is

$$f(x) = x' + b = 0 \tag{9.4}$$

A point source (vectors) xj and associated categories yj make up the training data. The $xj=Rd$ and the $yj=1$ for some dimension d, where Rd is a real value and b is a real number. As indicated in the diagram above, the ideal hyperplane for an SVM would be the one with the biggest margin between the two classes. With no inside data points, the margin is the slab's largest width parallel to the hyperplane. The support vectors are the number of observations closest to the Euclidean distance, which is located on the slab's edge (Figure 9.7).

The picture below illustrates these definitions, with+signifying type 1 pieces of data and –1 indicating type –1 data points. The dual quadratic programming problem is easier to solve in terms of computation. Accept Lagrange i multipliers compounded by each constraint, and remove them from the objective function to get the two alternatives.

In general, the largest concentration of i is zero. The non-zero i in the dual problem solution, as indicated in Eq. (9.1), defines the hyperplane as the sum of iy_ix_i. The data points x_i that correspond to non-zero i are the support vectors. SVMs define a decision boundary

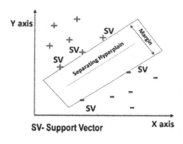

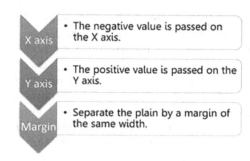

FIGURE 9.7
Support vector hyperplane.

and a maximum margin that divides nearly all points into two groups. Maximal margin algorithms have been replaced by support vector machines. Its most significant benefit is that it can use kernel functions to establish both a linear and non-linear decision boundary. This makes it better suited to real-world challenges where data aren't always perfectly separable by a straight line. The margin's position is determined by the vectors nearest to the decision boundary. As a result, the vectors on top of the margin are called support vectors. A margin is built around the decision boundary to address this issue. This margin's purpose is to keep the vectors as far away from the decision boundary as feasible. The idea is that having a buffer increases confidence in forecasts.

There is less ambiguity during classification since the vectors are at least the length of the margin away from the decision border. When sigma is little and the parameter is large, the SVM produces the best results. The regularized logistic regression, on the other hand, outperforms the generalized logistic regression. The regularization parameter "lambda" is to blame. The efficiency of logistic regression improves as the lambda value is lowered. However, the greatest results are obtained when lambda is set to a specific number (in our example, 0.0001). SVM (both logistic and random) efficiency improves as the number of rounds of categorizing the dataset increases while training duration increases.

9.5.3 DT Classifier

The classification algorithms category includes the DT algorithm. Regression and classification challenges can both be solved using the decision tree method. The purpose of utilizing a decision tree is to develop a training model that uses fundamental selection rules learned from training information to forecast the target variable's values. Learning and prediction are both included in the classification stage. The data gathered during the learning phase are used to construct the model. In the classification stage, the model is used to forecast the response for the available dataset. In DT, there are two types of variables: continuous and categorical variables.

- Continuous variable: *has a continuous target variable.*
- Categorical variable: *with a categorical target variable.*

Nodes and branches make up the tree. The subset specifies a value for each node, and each node represents features in a to-be-categorized category. A binary tree topology is a special database design categorization with at most two offspring for every connection point. The two child nodes are the left and right child nodes. When developing a DT, the most essential thing to remember is to select the best characters from the dataset's whole features list for the root node and subnodes.

The attribute selection measure is a method for selecting the finest qualities, and they are capable of handling multi-dimensional data. When compared to other algorithms, decision tree classifiers have high efficiency, and the hidden nodes may be seen using the leaf node. If the target attribute is a continuous variable, a regression model is created. The anticipated value of a target attribute while using a classification tree to forecast is the mid-point of the target attribute of a row that falls into a terminal (leaf) node of the tree. The ascribed objective is categorical; a classification model is constructed. Using a classification tree to forecast the value (category) of the target attribute, walk through the tree using the values of the predictor attributes until you reach a terminal (leaf) node, and then predict the category given for that connection point.

The following are the steps to classify DT algorithms:

i. First, at the root of the tree, keep the best feature of the input characteristics.

ii. Divide the training dataset into subsections next.

iii. Similar split subsets can be constructed by populating each with data that have a comparable input property value.

iv. Now repeat Steps 1–3 on each subgroup until the tree's leaf component is recognized in each branch.

The classification of traditional and correct diagnoses is investigated using decision tree classification (Figure 9.8). The classic approach to cancer diagnosis follows the standard

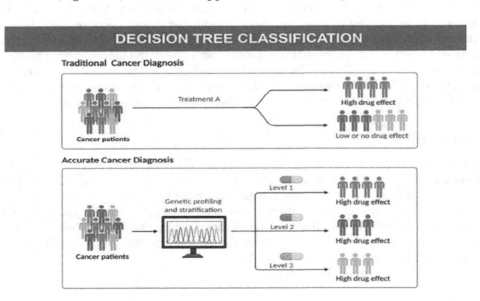

FIGURE 9.8
DT classification.

treatment methodology, where medications are given to patients according to a set of criteria such as low, medium, and high. This procedure and methodology do not result in an accurate diagnosis. Because of the lack of foresight in diagnosis, the risk factor is increasing. To overcome this practice, a structural procedure is implemented in order to optimize future prediction with technology. To determine an accuracy of tumor molecular classification and the subsequent estimation of datasets that help with cancer diagnosis, the patient's dataset is treated with a machine learning algorithm. There are three main stages and diagnoses for tumors: level 1, level 2, and level 3. The first-level therapy is offered to patients who have serious pharmacological side effects. Since the cancer is now in the early stages, it may be possible to recover from regression diagnosis and cure cancer altogether. In level 2, the tumor diagnosis curability is risky, but with a definitive diagnosis, approximately 60%–70% of people can try to conquer the tumor cell. Level 3 is the final stage of the disease; it cannot be cured, but patients can live with severe drug side effects.

9.6 Performance Evaluation

A formal and productive approach to measuring the efficiency of an algorithm and the result of a prediction is termed performance evaluation. The result performance is analyzed with sample datasets based on the algorithm analysis. The dataset was gathered for research purposes from Kaggle [25]. The effectiveness of the suggested classifier has been assessed using various metrics such as accuracy, precision, specificity, and F1-score.

Accuracy is the percentage of correctly classified values [26], and it is expressed as

$$Accuracy = \frac{True\ positive + True\ negative}{True\ positive + True\ negative + False\ positive + False\ negative} \tag{9.5}$$

Precision is used to calculate the model's ability to correctly categorize positive values [27]. The total number of anticipated positive values is divided by the number of true positives to arrive at this figure.

$$Precision = \frac{True\ positive}{True\ positive + False\ positive} \tag{9.6}$$

The model's recall is used to determine how effectively it can predict positive values [28]. By dividing the total number of real positives by the total number of genuine positives, the true-positive rate is derived.

$$Recall = \frac{True\ positive}{True\ positive + False\ negative} \tag{9.7}$$

The *F1*-score is the modulation index of recall and precision [29]. It comes in handy when both precision and memory are required.

$$F1\text{-}Score = \frac{2 * Precision * Recall}{Precision + Recall} \tag{9.8}$$

Binary categorization can result in four different outcomes:

- True negatives are those that were correctly predicted (zeros).
- Positives that were correctly predicted are known as true positives (ones).
- False negatives are negatives that were predicted inaccurately (zeros).
- False positives are positives that were predicted incorrectly (ones).

In binary categorization, classification accuracy is important.

- The proportion of positive instances to the overall number of true and false positives is the positive predictive value.
- The proportion of negative cases toward the overall number of true and false negatives is the negative predictive value.
- The ratio of true positives to actual positives is known as sensitivity.
- The specificity is defined as the ratio of genuine to actual negatives.

9.6.1 Result Analysis of LR Classifier

Any data collection can be evaluated using logistic regression. When the variable is categorical, it calculates the statistical approach for eliminating binary classes. The term binary refers to the fact that there are two possible classes, such as binary classes (0 and 1). Both classification and regression are done using logistic regression. It calculates the likelihood of an event occurring (Figure 9.9).

The values are grouped depending on the dataset's cancer prediction. Two markers of possible tumor progression are the proportion of actual and the probability of false alarm. If the positive rate is larger, there's a probability the cancer cell will get bigger and bigger. The rate of spread is controlled, and cancer is controlled when the false-positive rate is decreased.

9.6.1.1 Confusion Matrix for LR Classifier

A confusion matrix is a tabular pattern that aids in visualizing the various outcomes of a classification problem's forecast and results. True and false values are used to classify the items. The results of a machine learning classification task containing two maybe more output classes. In this table, there are four distinct projections and actual values.

- The number of times our actual positive values are equal to the forecasted positive values. You were correct in predicting a positive value.
- The number of times our model incorrectly predicts negative values as positives. You expected a negative number, but it turned out to be positive.
- **True Negative:** The number of times our real negative values and forecasted negative values are the same. You correctly predicted a negative value, and it is correct.
- The number of times our model incorrectly predicts negative values as positives. You expected a negative number, but it turned out to be positive.

	precision	recall	f1-score	support
Malignant(Class 0)	0.97	0.99	0.98	72
Benign (Class 1)	0.98	0.95	0.96	42
accuracy			0.97	114
macro avg	0.97	0.97	0.97	114
weighted avg	0.97	0.97	0.97	114

`<sklearn.metrics._plot.roc_curve.RocCurveDisplay at 0x15aeeede400>`

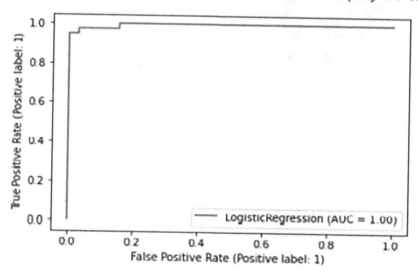

FIGURE 9.9
LR analysis.

The confusion matrix's output is determined by its accuracy, precision, recall, and *F*1-score. A binary classification improves the output of future tumor spread. These figures aid in the diagnosis and treatment of cancer. True-positive value is 71, whereas true-negative value is 40. There will be changes in future treatment based on these data (Figure 9.10).

9.6.2 Analysis of SVM Classifier

The goal of the SVM method is to find the best vector or determination boundary for categorizing n-dimensional data into categories such that subsequent data points can be easily placed in the relevant category. In SVM, two parameters are used to determine the appropriate threshold values for estimating how well a natural occurrence is a "benign" (1) or a "malignant" (0). A specific percentage of dataset rows will be accurately categorized, while another percentage will be misclassified. The ratio of positive instances to true negatives is measured using two metrics: malignant and benign. To put it another way, sensitivity is the percentage of 1s that are categorized as 1s using that specific decision threshold, also known as the true positive. Predictability, on the other hand, is the percentage of 0s something that has previously been predicted as 0s, also known as the genuine negative rate.

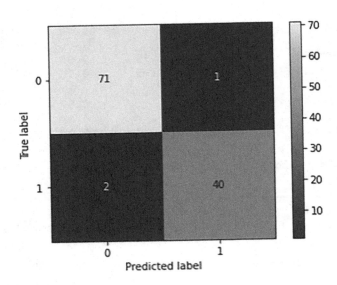

FIGURE 9.10
Confusion matrix for LR classifier.

These are expressed quantitatively as

- Benign = (number of true recognized 1s)/(total number of observed 1s).
- Malignant = (number of correctly recognized 0s)/(total number of observed 0s).

The malignant and benign classes are used to calculate the accuracy of tumor cells (Figure 9.11). The level of metrics is determined by precision, recall, F1-score, and support. Precision is a deterministic value that tumor cells have when they can develop deeper. The current status of the tumor cells is measured by the recall. The genuine positive and true-negative score of the classifications are measured by the F1-score. The harmonic F1-score considers true positive and false positive with equal weight and is a special type of average or special type of mean. The SVM algorithm calculates the accuracy of tumor diagnosis based on the measure.

9.6.2.1 Confusion Matrix for SVM Classifier

The confusion matrix is a table that shows whether a classifier performs on a set of test data that have established real values. The forecast's findings are summed up in a confusion matrix. In number count, a confusion matrix displays both right and wrong values. It assists us in creating a decent data visualization. It reveals not only the number of errors produced by a classifier, but also the sorts of errors made. The model forecasts true or false values (Figure 9.12).

The true label and prediction label are the two variables that show the tumor accuracy outcome. "Yes" and "no" are the two potential predicted classes. The forecasting presence of a disease "yes" would indicate that they have the condition, while "no" would indicate that they do not. A maximum of predictions were being tested for the presence of that disease. The classifier correctly predicted "yes" maximum and "no" occasions minimum. In actuality, a maximum of the participants in the study has the condition, while the minimum does not have disease overspread.

	precision	recall	f1-score	support
Malignant(Class 0)	0.97	0.99	0.98	72
Benign (Class 1)	0.98	0.95	0.96	42
accuracy			0.97	114
macro avg	0.97	0.97	0.97	114
weighted avg	0.97	0.97	0.97	114

: `<sklearn.metrics._plot.roc_curve.RocCurveDisplay at 0x29674440fd0>`

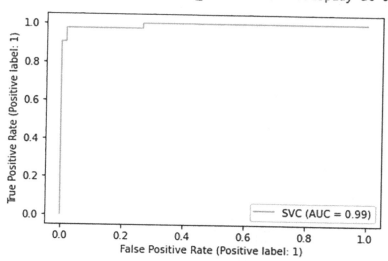

FIGURE 9.11
Analysis of SVM.

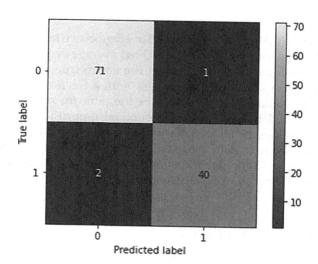

FIGURE 9.12
SVM confusion matrix.

	precision	recall	f1-score	support
0	0.94	0.94	0.94	72
1	0.90	0.90	0.90	42
accuracy			0.93	114
macro avg	0.92	0.92	0.92	114
weighted avg	0.93	0.93	0.93	114

`<sklearn.metrics._plot.roc_curve.RocCurveDisplay at 0x15aeef59700>`

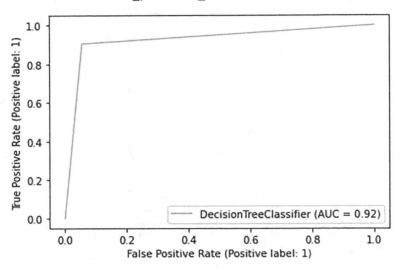

FIGURE 9.13
Analysis of DT classifier.

9.6.3 Analysis of DT Classifier

The sampling mean (estimated the unweighted average every labeled) and also the linear combination (averaged nearly the support-weighted average every label) are two types of averages. The micro-average (averaged almost the entire true positives, false negatives, and false positives) is just shown for multi-class with a fraction of categories because it correlates dependability and would otherwise be the same for all metrics.

Figure 9.13 predicts the dataset's true-positive and false-positive values. Precision, recall, F1-score, and support values are all important predicted by the classes 0 and 1. The macro-average and weighted average data are used to estimate tumor accuracy. They differ in terms of algorithm testing. In comparison with the algorithm, the decision tree defers such that the value of prediction can examine the accuracy of tumor spread and aid in future prediction.

9.6.3.1 Confusion Matrix for DT Classifier

A confusion matrix, also known as an error matrix, would be a table model that allows you to evaluate the performance of a training data system. The samples in a binary classifier are represented in each column of the matrix, while the occurrences in an actual class are

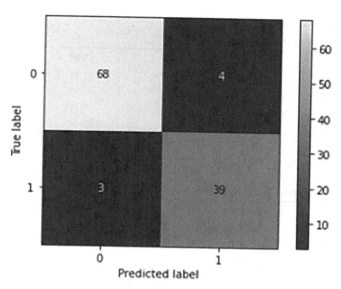

FIGURE 9.14
Confusion matrix of DT classifier.

represented in each row. Any non-zero values outside the diagonal can easily visualize faults because all right estimates are placed along the table's diagonal. A confusion matrix of size N×N coupled with classifiers shows the expected and observed categorization, where N represents the number of available classes. Using 0s and 1s, every classification is done automatically. Using accuracy, macro-average, and weighted average, a confusion matrix was trained using various metrics. The inaccuracy and carelessness of curability are examined, while future desired recognition will be monitored for subsequent therapy (Figure 9.14).

The measures of confusion DT are determined by the true label and prediction label. Examined the DT algorithm which predicts the future value of abnormal cell growth in the body. The two classes are classified with binary values as true or false. With that value, the confusion matrix is determined. The confusion matrix is drawn with a 2×2 matrix. The distribution of cell TP value shows the value of 68. The prediction rate of spreading is 39. The rate of curability and disorder of unrestricted cell proliferation are shown.

9.7 Comparative Analysis of Various Classifiers

The comparative analysis is based on specialized ML techniques such as LR, SVM, and DT in terms of accuracy, precision, recall, and F1-score. The input datasets are used to calculate each metric. The accuracy and outcome of a perfect algorithm are determined by algorithm analysis. It is indeed a team effort to figure out which algorithm is the best fit for each other. The following algorithms are used to forecast breast cancer using a sample dataset, which aids in the prediction of relevant data for examining the outcome value of future predictions. The data are shown graphically in graphs that depict tumor cell proliferation and characterization. The confusion matrix is compared with the prediction of true

TABLE 9.1

Malignant: Class 0 – Comparative Analysis

Metric	LR	SVM	DT
Accuracy	97.02	97.34	94.00
Precision	97.24	97.00	94.26
Recall	99.72	99.18	94.78
F1-score	98.68	98.32	94.24

TABLE 9.2

Malignant: Class 1 – Comparative Analysis

Metric	LR	SVM	DT
Accuracy	97.56	97.28	93.00
Precision	98.00	98.88	90.86
Recall	95.78	95.64	90.24
F1-score	96.90	96.10	90.72

or false values, yielding the F1-score with "yes" or "no" values, corresponding to "true" or "false" values (Tables 9.1 and 9.2).

True predictions in Class 1 are investigated. By comparing the two classes, it is simple to predict the disease's future recovery at various phases. Class 0 defines the false value of prediction, while Class 1 determines the genuine value. A comparison of the methods of logistic regression and SVM reveals that the mapping of future prediction is similar, where DT differentiates between logistic and SVM classifiers.

9.8 Conclusions

This chapter discussed cancer in depth, including abnormal cell proliferation and distribution in the body and organs. Cancer is the world's second-largest and fastest-growing disease. There are no full-spectrum treatments for tumors. In the world, more research and prevention approaches are continually being developed. This case study establishes the future prediction for tumor cells, followed by machine testing algorithm-based prediction. Future abnormal cell proliferation is predicted with the use of technology using various classifiers and algorithms. Sample data from the patient's medical history are utilized to assess future treatment options in the early stages of cancer detection. This shortens the therapy period and limits the patient's potential to heal. To overcome such difficulties, the medical industry has enrolled in new technology that aids in the resolution of man-made problems. To avoid such issues, medical technology has brought new future using machine learning algorithms. Breast cancer is analyzed utilizing a sample dataset and several classifiers such as LR, SVM, and DT as an example of technology. To determine the correctness of an algorithm, various metrics are utilized. As a comparative analysis of classifiers, LR and SVM replicate the same outcome of future prediction, indicating that they are effective in predicting tumor cell proliferation in future.

References

[1] T. Liu, J. Huang, T. Liao, R. Pu, A hybrid deep learning model for predicting molecular subtypes of human breast cancer using multimodal data, IRBM, 43, 62–74, 2022.

[2] S. Lavanya, et al., A tuned classification approach for efficient heterogeneous fault diagnosis in IoT-enabled WSN applications, *Measurement*, 183, 109771:1–16, 2021.

[3] C.A. Clarke, J.J. Earl Hubbell, Multi-cancer early detection: A new paradigm for reducing cancer-specific and all-cause mortality, *Cancer Cell*, 39, 447–448, 2021.

[4] K.B. Bhaskar, et al., An energy-efficient blockchain approach for secure communication in IoT-enabled electric vehicles, *International Journal of Communication System*, 35, e5189:1–25, 2022.

[5] A. Prasanth, S. Jayachitra, A novel multi-objective optimization strategy for enhancing quality of service in IoT-enabled WSN applications, *Peer-to-Peer Networking and Applications*, 13, 1905–1920, 2020.

[6] S. Dhiviya, S. Malathy, D.R. Kumar, Internet of things (IoT) elements, trends and applications. *Journal of Computational and Theoretical Nanoscience*, 15(5), 1639–1643, 2018.

[7] M. Lua, Z. Fand, B. Xu, Using machine learning to predict ovarian cancer, *International Journal of Medical Informatics*, 141, 104195:1–15, 2020.

[8] D. Lakshmi, S.R. Gurrela, M. Kuncharam, A comparative study on breast cancer tissues using conventional and modern machine learning models, In: Satapathy, S.C., Bhateja, V., Favorskaya, M.N., Adilakshmi, T. (eds) *Smart Computing Techniques and Applications. Smart Innovation, Systems and Technologies*, Springer, Singapore, 225, 693–699, 2021.

[9] E. Hubbell, C.A. Clarke, A.M. Aravanis, C.D. Berg. Modeled reductions in late-stage cancer with a multi-cancer early detection test, *Cancer Epidemiology, Biomarkers & Prevention*, 30, 460–468, 2021.

[10] D. Sun, A. Li, B. Tang, M. Wang, Integrating genomic data and pathological images to effectively predict breast cancer clinical outcome, *Computer Methods and Programs in Biomedicine*, 161, 45–53, 2018.

[11] S.R. Jena, R. Shanmugam, R.K. Dhanaraj, K. Saini. Recent advances and future research directions in edge cloud framework, *International Journal of Engineering and Advanced Technology*, 9(2), 2019.

[12] S. Sharma, R. Kamala, D. Nair, T. Raju Ragavendra, Round cell tumors: Classification and immunohistochemistry, *Indian Journal of Medical and Paediatric Oncology*, 38(3): 349–353, 2017.

[13] M. Cristofanilli, G. Thomas Budd, M.J. Ellis, Circulating tumor cells, disease progression, and survival in metastatic breast cancer, *The New England Journal of Medicine*, 351(8), 781–791, 2004.

[14] J.G. Liao, K.-V. Chin. Logistic regression for disease classification using microarray data: Model selection in a large p and small n case, *Bioinformatics*, 23(15), 1945–1956, 2007.

[15] W. Shao, T. Wang, L. Sun, Multi-task multi-modal learning for joint diagnosis and prognosis of human cancers, *Medical Image Analysis*, 65, 101795:1–16, 2020.

[16] Y.H. semann, J.B. Geigl, F. Schubert, P. Musiani. Systemic spread is an early step in breast cancer, *Cancer Cell*, 13(1), 58–68, 2008.

[17] B.N. Patel, S.G. Prajapati, K.I. Lakhtaria, Efficient classification of data using decision tree, *Bonfring International Journal of Data Mining*, 2(1), 6–12, 2012.

[18] M. Rana, P. Chandorkar, A. Dsouza, N. Kazi Breast cancer diagnosis and recurrence prediction using machine learning techniques, *IJRET: International Journal of Research in Engineering and Technology*, 4, 372–376, 2015.

[19] F.B. Aissa, M. Sakkari, R. Ejbali, M. Zaied, Unsupervised features extraction using a multi-view self-organizing map for image classification, *Proceedings in IEEE/ACS 14th International Conference on Computer Systems and Applications*, 196–201, 2017.

[20] O. Kott, D. Linsley, A. Amin, Development of a deep learning algorithm for the histopathologic diagnosis and Gleason grading of prostate cancer biopsies: A pilot study, *European Urology Focus*, 7, 347–351, 2019.

[21] S.A.A. Ismael, A. Mohammed, H. Hefny, An enhanced deep learning approach for brain cancer MRI images classification using residual networks, *Artificial Intelligence in Medicine*, 102, 101779:1–21, 2020.

[22] C.-Y. Joanne Peng, K.L. Lee, G.M. Ingersoll, An introduction to logistic regression analysis and reporting, *The Journal of Educational Research*, 96, 1–17, 2002.

[23] A. Elsayad, Diagnosis of breast cancer using decision tree models and SVM, *International Research Journal of Engineering and Technology*, 5, 2845–2848, 2018.

[24] H. Rajaguru, S.R. Sannasi Chakravarthy, Analysis of decision tree and k-nearest neighbor algorithm in the classification of breast cancer, *Asian Pacific Journal of Cancer Prevention*, 20, 3777–3781, 2019.

[25] https://www.kaggle.com/datasets/rishidamarla/cancer-patients-data.

[26] A. Prasanth, Certain investigations on energy-efficient fault detection and recovery management in underwater wireless sensor networks, *Journal of Circuits, Systems and Computers*, 30, 2150137:1–20, 2021.

[27] S. Kalli, et al., An effective motion object detection using adaptive background modeling mechanism in video surveillance system, *Journal of Intelligent & Fuzzy Systems*, 41, 1777–1789, 2021.

[28] S. Irfan, R.K. Dhanaraj, BeeRank: A heuristic ranking model to optimize the retrieval process, *International Journal of Swarm Intelligence Research (IJSIR)*, 12(2), 39–56, 2021.

[29] J. Sekar, P. Aruchamy, An efficient clinical support system for heart disease prediction using TANFIS classifier, *Computational Intelligence*, 38, 610–640, 2022.

10

A Deep Exploration of Imaging Diagnosis Approaches for IoMT-Based Coronavirus Disease of 2019 Diagnosis System – A Case Study

Preethi Sambandam Raju, Revathi Arumugam Rajendran, and Murugan Mahalingam

SRM Valliammai Engineering College

CONTENTS

10.1 Introduction

The coronavirus disease of 2019 (COVID-19) is a pandemic and serious disaster from December 2019. As on April 12, 2020, 9:42 am GST, nearly 17,80,717 persons were affected with 1,08,837 deaths reported across the world. By April 23, 2020, 3:31 pm GST, about 26,56,622 persons were reported with infectious coronavirus disease and 1,85,166 deaths were reported [1]. Thus, in the span of 11 days, the infection had spread to nearly two and a half lakhs people and the death rate had increased by approximately 77,000. So far, the only way to flatten the curve of coronavirus disease reports is to detect persons with symptoms

DOI: 10.1201/9781003256243-10

and isolate them. Hence, the coronavirus disease symptom imaging device based on the Internet of things and artificial intelligence is built.

The symptoms of coronavirus disease include coughing, fever, body aches, tiredness, and difficulty breathing. In an outbreak of coronavirus disease symptom imaging device, sound classification-based cough detection and elevated body temperature detection with coughing action using thermal camera surveillance are implemented. Internet of things (IoT) is generally used for many automation applications [2–5]; when combined with artificial intelligence, it can be used as an automatic intelligent detection system. IoT and video surveillance (VS) are merged to give VS-IoT and are implemented for fast and reliable surveillance. The same technology is used with the required alteration for thermal camera surveillance.

The main contributions include (a) imaging techniques, (b) coronavirus disease of 2019 features to be pondered, (c) algorithms deployed for imaging techniques, (d) coronavirus disease of 2019 dataset details, (e) discussion of recent related works, (f) architecture and implementation of the outbreak of coronavirus disease symptom imaging device, and (g) details of the implemented indication-notifying algorithm and conclusions of the work with a future focus.

10.2 Backgrounds of Imaging Techniques

There are three imaging techniques as shown in Figure 10.1, namely positron emission tomography (PET), ultrasonography, and magnetic resonance imaging, which help to view the coronavirus disease-infected regions (Figure 10.2).

10.2.1 PET

Positron emission tomography (PET) exists as a delicate, but aggressive picturing procedure. It plays a significant part in assessing inflammatory pulmonary diseases and infectious pulmonary diseases. Further, it also helps in observing disease development and treatment consequences and thus helps in enlightening patient management. Fluorodeoxyglucose is a radiopharmaceutical put into use within the medicinal picturing modality PET. FDG uptake mentions the quantity of radiotracer acceptance. F-FDG PET/CT examinations reveal coronavirus disease-linked pneumonia in asymptomatic persons during the COVID-19 pandemic.

F-FDG PET/CT can show a supplementary investigative part in coronavirus disease of 2019, particularly put into the initial phase, once the difference analysis is found to be problematic. In the middle of the single situation statement, an increased acceptance of

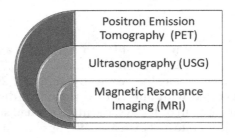

FIGURE 10.1
Types of imaging techniques.

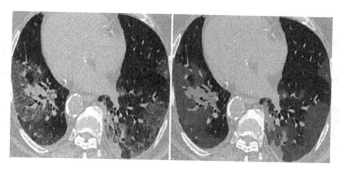

FIGURE 10.2
Illustration of coronavirus disease of 2019-infected area.

FDG initiated inside the lymphatic nodules and bony kernel through fluorodeoxyglucose PET/CT valuation. This fluorodeoxyglucose PET/CT is an additional multifaceted examination compared to trunk CT, and an extended examination duration for fluorodeoxyglucose PET/CT inspections might upsurge the danger of disease broadcast.

10.2.2 Lung Ultrasound

The habit of ultrasonography (USG) in the crisis sector, life-threatening care, and cardiac caution entities is flatteringly prevalent. This imaging technique is effortlessly obtainable at the side of the bed, existent period and void of radioactivity menaces present in comparison with the conservative picturing procedure of lung in depreciatively ailment prevailing persons. Lung ultrasound accompanies conservative valuation approaches and further picturing procedures of the lung [6,7]. The emergent application of LUS in dissimilar locales has directed modification in methodology and classification.

The authors of [8] stated that pulmo-ultrasonography is able to deliver consequences similar to trunk CT for the assessment of coronavirus disease pneumonia. On behalf of expecting females by way of supposed coronavirus disease, trunk CT inspection must be circumvented as ample as conceivable payable to the extraordinary emission quantity danger to the growing fetus. By means of a substitute, the authors of [9] suggested that obstetricians and gynecologists execute trunk investigations via LUS.

10.2.3 MRI

An MRI scan habits a big magnet, radio-waves, and a computer to generate a comprehensive, sectional copy of interior body parts and structures [10]. The scanner itself characteristically looks like a huge pipe with a bench in the center, letting the patient to shot in. An MRI scan varies from computer-assisted tomography images and X-rays, as it does not use possibly damaging ionizing energy. However, it is not usually pragmatic to coronavirus disease analysis owing to a comparatively extensive skimming period and great price associated with computer-assisted tomography and LUS. Nevertheless, the non-intervention MRI might be implemented with the assessment of coronavirus disease in kids and expectant women.

The contagion of SARS-CoV-2 is mostly dispersed in the trunk of the human body; nevertheless, three negligibly aggressive dissections presented by the side of the contamination similarly include the indemnities of receptacles, heart, liver, kidney, and further body part [11].

TABLE 10.1

Comparison of Imaging Techniques

PET	CT	MRI
The problems at the cellular level are identified.	It scans the body's internal organs and tissues.	It scans the body's internal organs and tissues.
The patient is injected with a liquid that contains radioactive substances.	3D picture of the injury is obtained.	More details are provided by MRI.
It uses less radiation compared to CT.	It contains radiation.	It has no radiation at all.

By the side of the RNA value, Zou et al. [12] acknowledged further body parts in danger, such as the heart, esophagus, kidney, bladder, and ileum, that are vulnerable to coronavirus disease contamination. A comparison of imaging techniques for COVID is tabulated in Table 10.1.

10.3 Features to Be Pondered of COVID

Representative picturing physiognomies of the trunk in coronavirus disease embrace lacerations by means of bilateral patchy shadowing, pulmonary fibrosis, multiple lesions, ground glass opacity (GGO), lung amalgamation, and crazy-paving pattern, as depicted in Figure 10.3. These picturing elucidations frolicked a crucial part not only in the analysis and management of coronavirus disease, but also in the perceiving of sickness development and the assessment of beneficial efficiency. The age-wise pattern is also detailed [13].

10.4 Algorithms Deployed for Imaging Techniques

Imaging techniques along with the support of algorithms help to detect the coronavirus disease of 2019 infections. Algorithms such as Visual Geometry Group (VGG) Net,

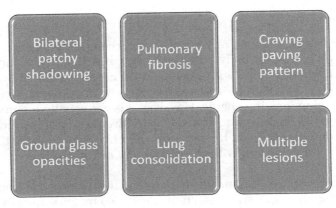

FIGURE 10.3

Imaging physical appearance of lung coronavirus disease.

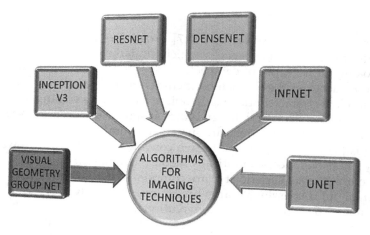

FIGURE 10.4
Algorithms deployed for imaging techniques.

Inception V3, ResNet, DenseNet, Inf-Net, UNet, and many machine learning techniques as depicted in Figure 10.4 help in coronavirus disease of 2019 identification.

10.4.1 Visual Geometry Group Net

VGG Net [14] from the Visual Geometry Group is uniquely present in the greatest prevalent deep CNN styles, which protected the first and second spots object localization besides ordering chores in ILSVRC 2014. In this design, the chief impression remained that the cumulative complexity of the CNN designs and substituting great quintessence by manifold slighter quintessence remained possibly extra precise through booming out computer visualization chores. VGG Net alternatives are cycled fairly expansively for numerous computer visualization chores aimed at mining deep picture sorts, on behalf of auxiliary handling, particularly in the medicinal imaging arena.

10.4.2 Inception V3 Designs

In InceptionV3 designs [15], the chief impression is to covenant with the problematic of dangerous erraticism in the site of the noticeable shares in the images under contemplation by allowing the network to enclose manifold diverse categories of kernels in the similar level.

10.4.2.1 ResNet

The important impression in ResNet designs presented in the work designed by He et al. [16] is in which piling up and about of convolutional as well as pooling coating layered zone one arranged topmost of one more, be able to source the system presentation toward impairment, in arrears on the way to tricky vanishing incline, consequently, toward pact using this, individuality time-saving method be able to be deployed, that can essentially hop any one or additional layers. The group coatings that comprise individuality influences are named an enduring chunk. The impression of totaling bounce acquaintances basically develops a purge of the great exercising blunder, which is characteristically witnessed in an alternative profound style.

10.4.2.2 DenseNet

The author projected the DenseNet style [17] progresses arranged ResNet designs by means of integrating profound influences, which fundamentally associate each layer with every other coating. It is caring of compactly associated architectures to safeguard that respective coating acquires the characteristic charts since apiece foregoing layer then permits on its individual characteristic map to apiece succeeding coating. An additional significant benefit of such a structural design is the capability to reprocess topographies, even though a small amount of restrictions are sustained overall.

CT imaging is principally nurtured to two convolutional stratums toward intangible extraordinary – perseverance, semantically feeble (i.e., small in number) sceneries [18,19]. In this, they complement a superiority care component toward obviously increasing the demonstration about dispassionate area limitations. Formerly, the lower value features $f2$ acquired remain nourished to three convolutional layers for mining the elevated level landscapes, which are there used for two determinations. Primarily, it consumes a parallel partial decoder (PPD) to collect the attribute characters and then produce a worldwide map.

10.4.2.3 Inf-Net

Presently, at hand is a very restricted quantity of CT images with separation clarifications; meanwhile, physically segmenting lung contamination constituencies is problematic and laborious, and the ailment is at a premature period of occurrence. To steadfastness this dispute, it requisite progress Inf-Net [20] expending a semi-supervised education stratagem, which influences an enormous numeral of unspecified CT imaginings to meritoriously supplement the preparation data repository. Presently, there is a precise degree of CT images with splitting up clarifications; meanwhile, physically sectoring body trunk contamination provinces are problematic and period – overwhelming, and the ailment is at a primary phase of the eruption. In the direction of steadfastness in this dispute, Inf-Net progresses by means of a partial supervising learning scheme, which influences a huge number of unspecified CT pictures to meritoriously supplement the drilling data repository.

10.4.2.4 UNet

Olag Ronneberger et al. introduced the UNet architecture [21] for biomedical image segmentation. The presented styles had two foremost fragments, which were found to be encoder along with decoder. The encoder is completely around the contract layers trailed by the pooling process. This stands to be used to excerpt the issues within the picture. The additional portion of decoder customs transferred convolution to license determination. This one stays once more an F.C linked stratums linkage-efficient CNN structural design baptized ShuffleNet, that is premeditated individually for moveable procedures with precise some degree of computing supremacy (e.g., 10–150 MFLOPs).

The innovative structural design makes the most of two novel procedures, pointwise collection involvedness and channel shambles, to prominently decrease reckoning cost while upholding accurateness. Experimentations taking place in ImageNet cataloguing besides MS COCO object recognition make obvious grander enactment about ShuffleNet in excess of additional arrangements, e.g., lesser top-1 fault than current MobileNet on ImageNet organization assignment, beneath the calculation economical of 40 MFLOPs. Taking place in an ARM-grounded mobile apparatus, ShuffleNet attains ~13× definite speedup above AlexNet although conserving analogous correctness.

10.5 Machine Learning Techniques

A contemporary examination of coronavirus disease projection/jeopardy forecast approaches described that the furthermost about them are prejudiced owing to either one of the two clarifications.

- At the outset, voluminous distributed educations absence quantifiable updating data, entailing which the classifications (tags) utilized on behalf of machine learning may be imprecise, since ill persons might progress undecorated difficulties consequently to the preliminary experimental bump utilized for ML.
- Second, numerous studies utilize the latter obtainable forecaster dimensions from automated well-being annals, reasonably compared to the forecaster standards assimilated at the period when the ideal is envisioned for usage.

Furthermore, numerous procedures would not comprise slight explanations of the education inhabitants, or else the envisioned utilization of the established copies, remain not illuminated unmistakably, or else stay with not comprehensively verified. At the supplementary circumstances, consideration standards are subjectively grouped, or else the experimentations on the bottom of the hyperparameter location remain not to be vigorous or are not conveyed. The aforementioned contemplations clue to the deduction that the furthermost of the obtainable approaches are below par designated in addition at in elevation jeopardy of prejudice, hovering apprehension in which their extrapolations might stand undependable the minute purposeful in medical repetition intended to advance a demanding and understandable risk calculation model that circumvents feebleness declared overhead.

An understandable machine learning conclusion organization grounded on additive trees, which are able to wherewithal doctors in the primary coronavirus disease menace valuation done a group of unpretentious besides human-predictable choice rubrics. Residual neural network (ResNet) remains to be a type of ANN of a genus that constructs on notions documented commencing on pyramid cells present within the cerebral cortex. ResNets concoct this by retaining hop influences, or time-saving methods to hurdle in excess of nearly some layers. Typical residual neural network [22] replicas remain engaged through dual- or triple-decker layer bounces that realize non-linearities (ReLU) in addition to bunch standardization in regarding. A supplementary heft matrix could be utilized to obtain the bounce masses; these replicas are recognized as Highway Net [23] replicas through numerous equivalent bounces stand denoted nearby means of DenseNet. In the milieu of ResNet, a non-ResNet could be demarcated such as a bare net.

A big inspiration for hopping above layers stays to evade the tricky of endangered gradients, through recycling stimulations commencing a preceding layer till the neighboring layer studies its masses. In the course of preparation, the masses adjust to silence the upstream layer, in addition to intensifying the previously omitted layer. In this humblest circumstance, solitary masses aimed at the adjoining layer's connotation are revised, by means of no unambiguous burdens aimed at the above layer. These mechanisms are finest once a solitary non-linear layer is treading above, or once the in-between layers are all undeviating.

10.6 Coronavirus Dataset

The details of various available data repositories of COVID along with the number of samples considered for COVID-19 cases, other respiratory diseases, and samples of normal healthy patients are tabulated in Table 10.2.

TABLE 10.2

Details of Dataset with Analyzed Patients' Details

Patient Details			
Number of Respiratory-Related Disease Samples	Number of Coronavirus Patient Samples	Number of Normal Condition Samples	Dataset Reference No. and Authors
178 Samples of pneumonia		247	[24] https://github.com/ieee8023/CoronaVirusDiseaseof2019-chestxraydataset.
76 Bacterial pneumonia samples 48 SARS samples	521	397	[25] https://www.sirm.org/en/category/articles/CoronaVirusDiseaseof2019-19-database/
–	106	100	[26] ChainZhttp://www.ChainZ.cn.
–	X-ray samples: 117; CT samples : 20	X-ray samples: 117; CT samples: 20	[27] https://medRxiv.org/abs/2020.03.20.20039834
–	1,262	1,230	[28] https://www.kaggle.com/c/rsna-pneumonia-detectionchallenge.
1,385 Samples of respiratory disorders	496	–	[29] Depeursinge et al. (2012)
98 Samples of lung cancer 397 Samples of other lung diseases	449	100	[30] http://medicalsegmentation.com/CoronaVirusDiseaseof201919/
397 Samples of respiratory diseases	349	–	[31] Datasethttps://github.com/UCSD-AI4H/CORONA VIRUS DIESEASE OF 2019-CT.
660 Samples of lung disorders	564	–	[32] https://medRxiv.org/abs/2020.04.13.20063941.
97 Samples of respiratory ailments	53	–	[33] https://arXiv:2003.09424.
229 Samples of breathing disorders	313	–	[34] Wang et al. (2020)
55	51	–	[35] https://medRxiv.org/abs/2020.02.25.20021568
413 Samples of lung ailments	723	19	[36] https://medRxiv.org/abs/2020.03.19.20039354.
439 Samples of respiratory ailments	413	–	[37] Pathak et al. (2020)
397 Samples of lung illnesses	460	–	[38] Polsinelli et al. (2004)
–	230	130	[39] Han et al. (2020)
1,695 Respiration illness samples	1,029	–	[40] Harmon et al. (2020)
224 Samples of influenza	219	175	[41] Xu et al. (2020)

(Continued)

TABLE 10.2 (*Continued*)

Details of Dataset with Analyzed Patients' Details

	Patient Details		
Number of Respiratory-Related Disease Samples	**Number of Coronavirus Patient Samples**	**Number of Normal Condition Samples**	**Dataset Reference No. and Authors**
4,106 Non-COVID-19 samples	1,266	–	[42] Wang et al. (2020)
76 Samples of bacterial pneumonia 48 SARS	521	391	[43] Hu et al. (2020)
665 Non-COVID-19 samples	521	–	[44] Bai et al. (2020)
1,027 Community-acquired pneumonia	1,495	–	[45] Kang et al. (2020)
1,027 Community-acquired pneumonia	1,658	–	[46] Shi et al. (2003)
101 Samples of bacterial pneumonia	88	–	[47] Fang et al. (2020)
–	230	100	[48] Han et al. (2020)
1,357 Samples of pneumonia 444 Samples of lung cancer	1,194	998	[49] Ko et al. (2020)
1,735 Samples of pneumonia	1,292	713	[50] Li et al. (2020)
6,871 Samples of pneumonia	354	8,566	[51] Pu et al. (2020)
–	23,812	–	[68] Mohammed et al. (2020)
–	131	–	[55] Zhu et al. (2020)
	1,014 Suspected		[52] Ai et al. (2020)
101	101	–	[53] Zhao et al. (2020)
205 Samples of viral pneumonia	219	–	[54] Bai et al. (2020)
84 Samples of pneumonia	32	–	[55] Zhu et al. (2020)
–	83	–	[56] Li et al. (2020)
82 Samples of respiratory ailments	51	–	[57] Chen et al. (2020)
29 Samples of pneumonia	46	–	[58] Fang et al. (2020)
55 Samples of viral pneumonia	44	–	[59] Wang et al. (2020)
224 Samples of viral pneumonia	110	175	[60] Xu et al. (2020)
413 Samples of other respiratory diseases	723	–	[61] Jin et al. (2020)
100 Samples of bacterial pneumonia	88	–	[62] Song et al. (2020)
260 Samples of other respiratory disorders	496	–	[63] Jin et al. (2020)
1,551 Samples of community-acquired pneumonia	468	1,303	[64] Li et al. (2020)
1,027 Samples of community-acquired pneumonia	1,658	–	[65] Shi et al. (2020)
4,106 Samples of lung cancer 342 Samples of pneumonia	924	–	[66] Wang et al. (2020)
151 Samples of lung disorders	45	–	[67] Shi et al. (2020)

10.7 Related Works of Coronavirus Disease of 2019 and IoT

Coronavirus disease of 2019 IoT-based smart helmet with mounted thermal imaging for temperature detection is built for close analysis of body temperature [68]. Many works related to IoT and AI are also being done [69–71]. In the smart helmet, coughing is not monitored, which is also a major symptom. FluSense [72] is a Raspberry Pi-based system for detection built using thermal imaging, microphone, and neural network engine with Raspberry Pi. In FluSense, both coughing and elevated body temperature are considered. Nevertheless, wearing a mask may reduce the coughing sound, and the coughing action is not considered. Additionally, the edge computation puts a constraint on computational power and memory that is ruled out by deploying a cloud-based nanonet. The nanonet also can reduce the latency caused by cloud computing because of its small size and transfer learning methodology.

10.7.1 Coronavirus Disease of 2019 Symptoms Imaging System

Coronavirus disease of 2019 symptom imaging device can be visualized to have two working models such as cough detection and elevated body temperature detection. The architecture diagram of the outbreak of the coronavirus disease of 2019 symptom imaging and alerting system is portrayed in Figure 10.5.

Cough detection is implemented with the microphone connected to Raspberry Pi. The sound entering into the microphone is converted to spectrograms, and then the received sound spectrogram is compared with specified abnormal cough spectrograms if the match is found; the Raspberry Pi waits for thermal camera surveillance detection results. The spectrum analyzer gives the sound signal in the frequency domain. The representation of sound can be done in logarithmic and linear frequency scales.

In Figure 10.6, the spectrum analyzer image in linear frequency scale of clearing throat, normal cough, abnormal cough, and abnormal cough wearing a mask is pictured. The light and dark blue-shaded regions denote frequencies with high intensities. Green-shaded regions show low-intensity frequencies. It can also be noted that the intensity of sound decreases due to mask-wearing. Thus, microphone-received sound spectrogram is

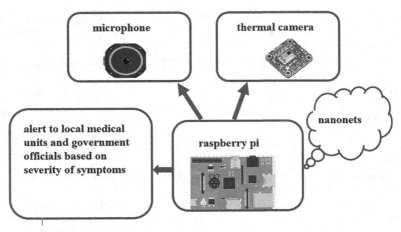

FIGURE 10.5
Architecture of coronavirus disease symptom imaging system.

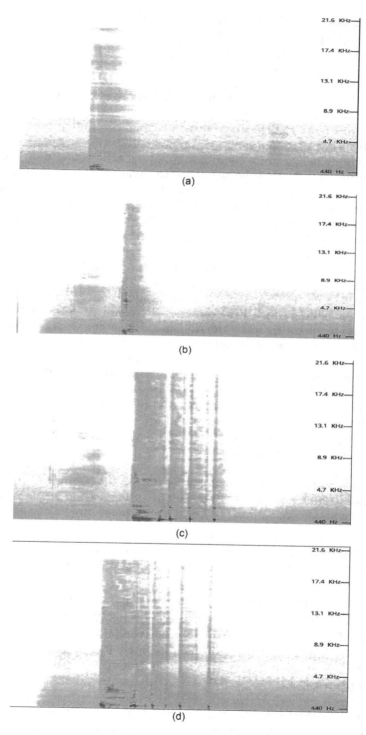

FIGURE 10.6

Spectrum analyzer output image of linear frequency scale: (a) clearing throat sound; (b) normal cough sound; (c) abnormal cough sound; (d) abnormal cough sound while wearing mask.

compared with abnormal cough and abnormal cough wearing mask spectrogram. Here, the abnormal cough spectrogram is utilized, as there is difficulty predicting the pattern of cough of coronavirus disease of 2019 only patients. This abnormal cough detects all coughs with breathing illnesses such as asthma, tuberculosis, and whooping sounds. Elevated body temperature detection is implemented with a thermal camera connected to Raspberry Pi; the thermal camera continuously clicks images and sends them to nanonets – nano-deep learning as a service platform [73].

Nanonets also can reduce the latency caused by cloud computing because of their small size and transfer learning methodology. Nanonets are tuned to detect three conditions such as elevated body temperature condition, coughing action condition, and normal conditions. For training and testing the nanonets, the available thermal camera images are used. Nanonets on sending the result to Raspberry Pi, based on the condition of cough detection the alert system works. The thermal images of the three varied conditions are revealed in Figure 10.6. Nanonets are a cloud platform with the input layer, pre-trained layer, nanonet layer, and output layer. Due to nanosize, the artificial intelligence platform can be easily deployed in Raspberry Pi.

The alerting system for the outbreak of coronavirus disease of 2019 symptom imaging device works with the indication-notifying algorithm. The Raspberry Pi gathers the result of two detection systems and implements the indication-notifying algorithm.

The indication-notifying algorithm combines the result of both cough detection unit and elevated body temperature using the AND logic to get the severity of coronavirus disease of 2019 symptoms. In case of coughing action or abnormal cough, the cough counter will help to enhance a better prediction in case of home surveillance. This counter should include both abnormal cough and cough actions together. The alerting system gives the GPS location to make the detection of persons with symptoms at ease. The alerting system action linked with the severity of coronavirus disease of 2019 symptoms that are obtained from cough detection and thermal image condition is described in Figure 10.7.

The benefits and shortcomings of individual picturing model in noticing lacerations could stand conceded as underneath:

- Trunk radiography displays no anomalies during the beginning-phase coronavirus disease of 2019; nonetheless, it cannot be utilized as an airing instrument for ill persons who are not endorsed toward for CT inspection;
- Trunk computerized tomography is a steadfast, firm, and useful picturing modality aimed at controlling the coronavirus disease of 2019, even though the monetary cost and continual acquaintance for ionizing contamination must remain deliberated;
- Intensely located body trunk lacerations could not be noticed by USG, and then they might be utilized in unsympathetically hostile ill persons to evaluate them;
- PET or CT delivers complementary evidence yonder that as long as by additional picturing models. That is found to be imperative toward adventure in the compensations of innumerable picturing models present in combat in contradiction of coronavirus diseases.

Approximately coronavirus disease of 2019 suffered patient's improvement to unadorned or life-threatening state. Ultrasonography is predominantly significant in scrutinizing disapprovingly sick patients and in deciding the course of handling, particularly

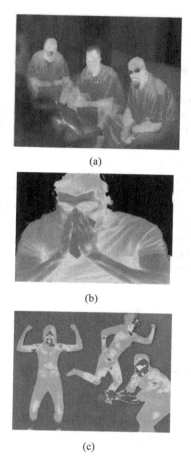

(a)

(b)

(c)

FIGURE 10.7
Available thermal image samples from Internet sources: (a) elevated body temperature condition; (b) coughing condition; (c) normal condition.

hemodynamic treatment. It emphasizes purposeful imaging more fairly than bodily imaging; in addition to this, solicitation may remain shortened by way of below.

- An USG inspection may stay utilized as a picturing model for speedy evaluation of disease. The dangerous session USG inspection remains optional for coronavirus disease of 2019 suffered by ill persons by means of indications of movement changeability, for rapid examination of probable impediments.
- Breathing sustenance is indispensable in unadorned and life-threatening circumstances, but the period of changing from non-invasive aeriation to invasive motorized aeriation is the foremost dispute amid medical physicians.
- The usage of moving urinary tubes and principal venous compression observation remains restricted through the surroundings of prevalent zones.
- Echocardiography can rapidly assess hemodynamics and classify if there are segmental wall gesture abnormalities or heart dysfunction that could lead to action.

10.8 Conclusions

The significant characteristic pieces to be deliberated and designated for coronavirus disease of 2019 analysis of chest scan are conferred in this chapter. Supplementary in this chapter, the features emphasizing the severity of COVID are intensely deliberated. The auxiliary usage of numerous artificial intelligence methods such as VGG-16, Inception V3, ResNet-50, DenseNet, Inf-Net, MSD-Net, ShuffleNet, CNN, UNet++, SVM, squeeze algorithm, BER algorithm, prior attention residual algorithm, weakly supervised deep learning, and decision fusion for imaging diagnosis converses. Alongside with this, the contrast and enactment of the above approaches are also meticulously projected.

Instruments and hardware positioned for imaging diagnosis of coronavirus disease of 2019 are elucidated. The extensive variability of datasets obtainable for COVID imaging examination of their possessions, benefits, and inadequacies discourse. An efficient IoT and artificial intelligence system for COVID symptoms detection with an alerting system is designed and implemented. Further, this is a low-cost and easily enactable system. It can help to detect persons with symptoms without the need for personnel to monitor and detect the cases in public and private places. The coughing action is tested through both sound and action, so all the doubtful cases will be covered. In the future, the spectrogram classification can be enhanced by COVID cough pattern.

References

[1] *World Meter: 'World Live Statistics of COVID19 Cases'*, https://www.worldometers.info/coronavirus/, accessed 12 April 2020 and 23 April 2020.
[2] Sambandam Raju, P., Mahalingam, M., Arumugam Rajendran, R. Design, implementation and power analysis of pervasive adaptive resourceful smart lighting and alerting devices in developing countries supporting incandescent and led light bulbs, *Sensors*, 19, 9, p. 2032, 2019, doi: 10.3390/s19092032.
[3] Sambandam Raju, P., Mahalingam, M., Arumugam Rajendran, R. Review of intellectual video surveillance through internet of things. In Peter, D., Alavi, A. H., Javadi, B., Fernandes, S.L, Eds. *Intelligent Data-Centric Systems, the Cognitive Approach in Cloud Computing and Internet of Things Technologies for Surveillance Tracking Systems*. Academic Press, 1st edn, 141–155, 2020.
[4] Bhaskar K.B, Prasanth A, Saranya P, An energy-efficient blockchain approach for secure communication in IoT-enabled electric vehicles, *International Journal of Communication System*, 35, e5189:1–25, 2022.
[5] Lavanya S, Prasanth A, Jayachitra S, A Tuned classification approach for efficient heterogeneous fault diagnosis in IoT-enabled WSN applications, *Measurement*, 183, 109771:1–22, 2021.
[6] Poggiali, E., et al., Can lung US help critical care clinicians in the early diagnosis of novel coronavirus (COVID-19) pneumonia? *Radiology*, 295, 3, E6, 2020, doi: 10.1148/radiol.2020200847.
[7] Kalafa, E., et al., Lung ultrasound and computed tomographic findings in pregnant woman with COVID-19, *Ultrasound in Obstetrics & Gynecology*, 55, 6, 835–837, 2020, doi: 10.1002/uog.22034.
[8] Peng, Q.Y., Wang, X. T., Zhang, L. N., Findings of lung ultrasonography of novel coronavirus pneumonia during the 2019–2020 epidemic, *Intensive Care Medicine*, 46, 5, 849–850, 2020, doi: 10.1007/s00134-020-05996-6.

[9] Moro, F., et al., How to perform lung ultrasound in pregnant women with suspected COVID-19, *Ultrasound in Obstetrics & Gynecology*, 55, 5, 593–598, 2020, doi: 10.1002/uog.22028.

[10] Jayachitra, S., Prasanth, A., An efficient clinical support system for heart disease prediction using TANFIS classifier, *Computational Intelligence*, 38, 610–640, 2022.

[11] Yao, A pathological report of three COVID-19 cases by minimal invasive autopsies, *Chinese Journal of Pathology*, 49, 5, 2020, doi: 10.3760/cma.j.cn112151-20200312-00193.

[12] Zou, X., Chen, K., Zou, J., Han, P., Hao, J., Han, Z., Single-cell RNA-seq data analysis on the receptor ACE2 expression reveals the potential risk of different human organs vulnerable to 2019-nCoV infection, *Frontiers of Medicine*, 14, 2, 185–192, 2020, doi: 10.1007/s11684-020-0754-0.

[13] Dong, D., et al., The role of imaging in the detection and management of COVID-19: A review, *IEEE Reviews in Biomedical Engineering*, 14, 16–29, 2021, doi: 10.1109/rbme.2020.2990959.

[14] Simonyan, K., Zisserman, A., Very deep convolutional networks for large-scale image recognition, *arXiv.org*, 2014. https://arxiv.org/abs/1409.1556.

[15] Szegedy, C., et al., Going deeper with convolutions, *arXiv.org*, 2014. https://arxiv.org/abs/1409.4842.

[16] He, K., Zhang, X., Ren, S., Sun, J., Deep residual learning for image recognition, *2016 IEEE Conference on Computer Vision and Pattern Recognition (CVPR)*, 2016, doi: 10.1109/cvpr.2016.90.

[17] Huang, G., Liu, Z., Van Der Maaten, L., Weinberger, K. Q., Densely connected convolutional networks, *2017 IEEE Conference on Computer Vision and Pattern Recognition (CVPR)*, 2017, doi: 10.1109/cvpr.2017.243.

[18] Ai, Correlation of chest CT and RT-PCR testing for coronavirus disease 2019 (COVID-19) in China: A report of 1014 cases, *Radiology*, 2019. https://pubs.rsna.org/doi/full/10.1148/radiol.2020200642.

[19] Ye, Z., Zhang, Y., Wang, Y., Huang, Z., Song, B., Chest CT manifestations of new coronavirus disease 2019 (COVID-19): a pictorial review, *European Radiology*, 30, 8, 4381–4389, 2020, doi: 10.1007/s00330-020-06801-0.

[20] Fan, D. P., et al., Inf-Net: Automatic COVID-19 lung infection segmentation from CT images, *IEEE Transactions on Medical Imaging*, 39, 8, 2626–2637, 2020, doi: 10.1109/tmi.2020.2996645.

[21] Weng, Y., Zhou, T., Li, Y., Qiu, X., NAS-Unet: Neural architecture search for medical image segmentation, *IEEE Access*, 7, 44247–44257, 2019, doi: 10.1109/ACCESS.2019.2908991.

[22] He, K., Zhang, X., Ren, S., Sun, J., Deep residual learning for image recognition, *2016 IEEE Conference on Computer Vision and Pattern Recognition (CVPR)*, 2016, doi: 10.1109/cvpr.2016.90.

[23] Zilly, J. G., Srivastava, R. K., Koutník, J., Schmidhuber, J., Recurrent highway networks, *PMLR*, 4189–4198, 2017. Available: https://proceedings.mlr.press/v70/zilly17a.html.

[24] Alom, M. Z., Rahman, M. M. S., Nasrin, M. S., Taha, T. M., Asari, V. K., COVID_MTNet: COVID-19 detection with multi-task deep learning approaches, *arXiv.org*, 2020. https://arxiv.org/abs/2004.03747.

[25] Hu, R., Ruan, G., Xiang, S., Huang, M., Liang, Q., Li, J., *Automated Diagnosis of COVID-19 Using Deep Learning and Data Augmentation on Chest CT*, 2020, doi: 10.1101/2020.04.24.20078998.

[26] Gozes, O., et al., Rapid AI development cycle for the coronavirus (COVID-19) pandemic: Initial results for automated detection & patient monitoring using deep learning CT image analysis, *arXiv.org*, 2020. https://arxiv.org/abs/2003.05037.

[27] Kassani, S. H., Kassasni, P. H., Wesolowski, M.J., Schneider, K. A., Deters, R., Automatic detection of coronavirus disease (COVID-19) in X-ray and CT images: A machine learning-based approach, *arXiv.org*, 2020. https://arxiv.org/abs/2004.10641.

[28] Jaiswal, N. G., Singh, D., Kumar, V., Kaur, M., *Classification of the COVID-19 Infected Patients Using DenseNet201 Based Deep Transfer Learning*, 2020. https://www.semanticscholar.org/paper/Classification-of-the-COVID-19-infected-patients-Jaiswal-Gianchandani/c01dcde84b6dc49af8c4b9388862383a30c3e807.

[29] Jin, C., et al. Development and evaluation of an AI system for COVID-19 diagnosis, *Nature Communications*, 11, Article number: 5088, 2020.

[30] Singh, D., Kumar, V., Kaur, M., Classification of COVID-19 patients from chest CT images using multi-objective differential evolution–based convolutional neural networks, *European Journal of Clinical Microbiology & Infectious Diseases*, 39, 7, 1379–1389, 2020, doi: 10.1007/s10096-020-03901-z.

[31] Amyar, Modzelewski, R., Li, H., Ruan, S., Multi-task deep learning based CT imaging analysis for COVID-19 pneumonia: Classification and segmentation, *Computers in Biology and Medicine*, 126, 104037, 2020, doi: 10.1016/j.compbiomed.2020.104037.

[32] Ahuja, S., Panigrahi, B. K., Dey, N., Rajinikanth, V., Gandhi, T. K., Deep transfer learning-based automated detection of COVID-19 from lung CT scan slices, *Applied Intelligence*, 51, 1, 571–585, 2020, doi: 10.1007/s10489-020-01826-w.

[33] Liu, B., et al. A fast online COVID-19 diagnostic system with chest CT scans. *Proceedings of KDD*. 2020, 2020.

[34] Barstugan, M., Ozkaya, U., Ozturk, S., Coronavirus (COVID-19) classification using CT images by machine learning methods, *arXiv.org*, 2020. https://arxiv.org/abs/2003.09424.

[35] Wang, X., et al., A weakly-supervised framework for COVID-19 classification and lesion localization from chest CT, *IEEE Transactions on Medical Imaging*, 39, 8, 2615–2625, 2020, doi: 10.1109/TMI.2020.2995965.

[36] Chen, J., et al., Deep learning-based model for detecting 2019 novel coronavirus pneumonia on high-resolution computed tomography, *Scientific Reports*, 10, 1, 2020, doi: 10.1038/s41598-020-76282-0.

[37] Wang, B., et al., AI-assisted CT imaging analysis for COVID-19 screening: Building and deploying a medical AI system, *Applied Soft Computing*, 98, 106897, 2021, doi: 10.1016/j.asoc.2020.106897.8.

[38] Pathak, Y., Shukla, P. K., Tiwari, A., Stalin, S., Singh, S., Shukla, P. K., Deep transfer learning based classification model for COVID-19 disease, *IRBM*, 2020, doi: 10.1016/j.irbm.2020.05.003.

[39] Polsinelli, M., Cinque, L., Placidi, G., A light CNN for detecting COVID-19 from CT scans of the chest, *Pattern Recognition Letters*, 140, 95–100, 2020, doi: 10.1016/j.patrec.2020.10.001.

[40] Han, Z., et al., Accurate screening of COVID-19 using attention-based deep 3D multiple instance learning, *IEEE Transactions on Medical Imaging*, 39, 8, pp. 2584–2594, 2020, doi: 10.1109/TMI.2020.2996256.

[41] Medical Segmentation, *COVID-19-Medical Segmentation*, 2020. http://medicalsegmentation.com/covid19/.

[42] Xu, X., et al., A deep learning system to screen novel coronavirus disease 2019 pneumonia, *Engineering*, 6, 10, 1122–1129, 2020, doi: 10.1016/j.eng.2020.04.010.

[43] Wang, S., et al., A fully automatic deep learning system for COVID-19 diagnostic and prognostic analysis, *European Respiratory Journal*, 56, 2, 2000775, 2020, doi: 10.1183/13993003.00775-2020.

[44] Hu, R., Ruan, G., Xiang, S., Huang, M., Liang, Q., Li, J., *Automated Diagnosis of COVID-19 Using Deep Learning and Data Augmentation on Chest CT*, 2020, doi: 10.1101/2020.04.24.20078998.

[45] Bai, H. X., et al., Artificial intelligence augmentation of radiologist performance in distinguishing COVID-19 from pneumonia of other origin at chest CT, *Radiology*, 296, 3, E156–E165, 2020.

[46] Kang, H., et al., Diagnosis of coronavirus disease 2019 (COVID-19) with structured latent multi-view representation learning, *IEEE Transactions on Medical Imaging*, 39, 8, 2606–2614, 2020, doi: 10.1109/tmi.2020.2992546.

[47] Shi, F., et al., Large-scale screening to distinguish between COVID-19 and community-acquired pneumonia using infection size-aware classification, *Physics in Medicine & Biology*, 66, 6, 065031, 2021, doi: 10.1088/1361-6560/abe838.

[48] Song, Y., et al., Deep learning enables accurate diagnosis of novel coronavirus (covid-19) with CT images, *IEEE/ACM Transactions on Computational Biology and Bioinformatics*, 18, 6, 2775–2780, 2021, doi: 10.1109/tcbb.2021.3065361.

[49] Ko, H., et al., COVID-19 pneumonia diagnosis using a simple 2D deep learning framework with a single chest CT image: Model development and validation, *Journal of Medical Internet Research*, 22, 6, e19569, 2020, doi: 10.2196/19569.

[50] Li, L., et al. Using artificial intelligence to detect COVID-19 and community-acquired pneumonia based on pulmonary CT: Evaluation of the diagnostic accuracy, *Radiology*, 296, 2, E65–E71, 2020.

[51] Ni, Q., et al., A deep learning approach to characterize 2019 coronavirus disease (COVID-19) pneumonia in chest CT images, *European Radiology*, 30, 12, 6517–6527, 2020, doi: 10.1007/s00330-020-07044-9.

[52] Korkmaz, I., Dikmen, N., Keleş, F. O., Bal, T., Chest CT in COVID-19 pneumonia: correlations of imaging findings in clinically suspected but repeatedly RT-PCR test-negative patients, *Egyptian Journal of Radiology and Nuclear Medicine*, 52, 1, 2021, doi: 10.1186/s43055-021-00481-6.

[53] Zhao, W., Zhong, Z., Xie, X., Yu, Q., Liu, J., Relation between chest CT findings and clinical conditions of coronavirus disease (COVID-19) pneumonia: A multicenter study, *American Journal of Roentgenology*, 214, 5, 1072–1077, 2020, doi: 10.2214/ajr.20.22976.

[54] Bai, H. X., B. Hsieh, Z. Xiong, K. Halsey, J. W. Choi, T. M. L. Tran, I. Pan, et al. Performance of radiologists in differentiating COVID-19 from non-COVID-19 viral pneumonia at chest CT, *Radiology*, 296, 2, E46–E54, 2020.

[55] Zhu, W., K. Xie, H. Lu, L. Xu, S. Zhou, S. Fang. Initial clinical features of suspected coronavirus disease 2019 in two emergency departments outside of Hubei, China, *Journal of Medical Virology*, 92, 9, 1525–1532, 2020.

[56] Li, K., et al., The clinical and chest CT features associated with severe and critical COVID-19 pneumonia, *Investigative Radiology*, 55, 6, 327–331, 2020, doi: 10.1097/rli.0000000000000672.

[57] Chen, J., et al., Deep learning-based model for detecting 2019 novel coronavirus pneumonia on high-resolution computed tomography, *Scientific Reports*, 10, 1, 2020, doi: 10.1038/s41598-020-76282-0.

[58] Fang, M., et al., CT radiomics can help screen the Coronavirus disease 2019 (COVID-19): A preliminary study, *Science China Information Sciences*, 63, 7, 2020, doi: 10.1007/s11432-020-2849-3.

[59] Wang, S., B. Kang, J. Ma, X. Zeng, M. Xiao, J. Guo, M. Cai, et al. A deep learning algorithm using CT images to screen for coronavirus disease (COVID-19), *European Radiology*, 2021, 1–9.

[60] Xu, X., et al., A deep learning system to screen novel coronavirus disease 2019 pneumonia, *Engineering*, 6, 10, 1122–1129, 2020, doi: 10.1016/j.eng.2020.04.010.

[61] Wang, B., et al., AI-assisted CT imaging analysis for COVID-19 screening: Building and deploying a medical AI system, *Applied Soft Computing*, 98, 106897, 2021, doi: 10.1016/j.asoc.2020.106897.

[62] Ma, J., et al., Toward data-efficient learning: A benchmark for COVID-19 CT lung and infection segmentation, *Medical Physics*, 48, 3, 1197–1210, 2021, doi: 10.1002/mp.14676.

[63] Song, Y., et al., Deep learning enables accurate diagnosis of novel coronavirus (COVID-19) with CT images, *IEEE/ACM Transactions on Computational Biology and Bioinformatics*, 18, 6, pp. 2775–2780, 2021, doi: 10.1109/tcbb.2021.3065361.

[64] Jin, C., et al., Development and evaluation of an AI system for COVID-19 diagnosis, 2020, doi: 10.1101/2020.03.20.20039834.

[65] Zheng, C., et al., Deep learning-based detection for COVID-19 from chest CT using weak label, 2020, doi: 10.1101/2020.03.12.20027185.

[66] Li, L., et al., Using artificial intelligence to detect COVID-19 and community-acquired pneumonia based on pulmonary CT: Evaluation of the diagnostic accuracy, *Radiology*, 296, 2, E65–E71, 2020, doi: 10.1148/radiol.2020200905.

[67] Shi, F., et al., Large-scale screening to distinguish between COVID-19 and community-acquired pneumonia using infection size-aware classification, *Physics in Medicine & Biology*, 66, 6, 065031, 2021, doi: 10.1088/1361-6560/abe838.

[68] Mohammed, M. N., H. Syamsudin, S. Al-Zubaidi, R. AKS, E. Yusuf. Novel COVID-19 detection and diagnosis system using IoT based smart helmet, *International Journal of Psychosocial Rehabilitation*, 24, 7, 2020, 2296–2303.

[69] Muhammad, L. J., Algehyne, E. A., Usman, S.S., Ahmad, A., Chakraborty, C., Mohammed, I. A., Supervised machine learning models for prediction of COVID-19 infection using epidemiology dataset, *SN Computer Science*, 2, 1, 2020, doi: 10.1007/s42979-020-00394-7.

[70] Jayachitra, S., Prasanth, A., Multi-feature analysis for automated brain stroke classification using weighted Gaussian Naïve Bayes classifier, *Journal of Circuits, Systems and Computers*, 30, 2150178:1–23, 2021.

[71] Sambandam Raju, P., Arumugam Rajendran, R., Mahalingam, M., Exploration of cough recognition technologies grounded on sensors and artificial intelligence, *Studies in Big Data*, 193–214, 2020, doi: 10.1007/978-981-15-8097-0_8.

[72] Hossain, F. A., Lover, A. A., Corey, G. A., Reich, N. G., Rahman, T., FluSense: A contactless syndromic surveillance platform for influenza-like illness in hospital waiting areas. *Proceedings of the ACM on Interactive, Mobile, Wearable and Ubiquitous Technologies*, 4, 1, 2020. https://dl.acm.org/doi/10.1145/3381014.

[73] *Thermal Imaging Condition Detection*, https://nanonets.com/, accessed on 2 April 2020.

Index

Printed in the United States
by Baker & Taylor Publisher Services